Scientific and Engineering Progress on Aluminum-Based Light-Weight Materials: Research Reports from the German Collaborative Research Center 692

Scientific and Engineering Progress on Aluminum-Based Light-Weight Materials: Research Reports from the German Collaborative Research Center 692

Special Issue Editor

Martin F.-X. Wagner

MDPI • Basel • Beijing • Wuhan • Barcelona • Belgrade

MDPI

Special Issue Editor
Martin F.-X. Wagner
TU Chemnitz, Institute of Materials Science and Engineering
Germany

Editorial Office
MDPI
St. Alban-Anlage 66
Basel, Switzerland

This is a reprint of articles from the Special Issue published online in the open access journal *Polymers* (ISSN 2073-4360) from 2017 to 2018 (available at: http://www.mdpi.com/journal/metals/special_issues/German_Collaborative_Research_Center_692)

For citation purposes, cite each article independently as indicated on the article page online and as indicated below:

LastName, A.A.; LastName, B.B.; LastName, C.C. Article Title. *Journal Name* **Year**, *Article Number*, Page Range.

ISBN 978-3-03897-196-2 (Pbk)
ISBN 978-3-03897-197-9 (PDF)

Contents

About the Special Issue Editor

Martin F.-X. Wagner, Univ. Prof. Dr.-Ing. habil., since 2010, holds the Chair of Materials Science at TU Chemnitz, Germany. After diploma studies in Mechanical Engineering at Ruhr-University Bochum, Prof. Wagner received a PhD in Materials Science in 2005. He spent one year as a Humboldt Scholar at the Ohio State University, USA, before returning to Bochum, where he led a Young Investigator's Research Group on Twinning, funded by the German Research Foundation's Emmy Noether Programme. His main research interests are the mechanical behavior of structural and functional materials across all length scales; experimental and theoretical aspects of twinning; light metals; severe plastic deformation; ultrafine-grained microstructures; ultra-high strength steels; and high strain rate testing. From 2014 to 2017, Prof. Wagner served as Speaker of the Collaborative Research Center SFB 692. He has received several awards, including the German Engineering Society's Golden Ring of Merit and the German Materials Society's DGM award.

Preface to "Scientific and Engineering Progress on Aluminum-Based Light-Weight Materials: Research Reports from the German Collaborative Research Center 692"

Academia and industry alike are faced with an ever-growing demand for energy-efficiency and reduced weights. Aluminum-based light-weight materials offer great potential for novel engineering applications, particularly when they are optimized to exhibit high strength and yet provide sufficient reliability. The last decade has thus seen substantial activity in the research fields of high-strength aluminum alloys and aluminum-based composite materials.

For twelve years, backed by solid funding from the German Research Foundation (Deutsche Forschungsgemeinschaft, DFG), scientists of the Collaborative Research Center "High-strength aluminum-based light-weight materials for safety components" (SFB 692) at TU Chemnitz, Germany, have contributed to this research area. Our research efforts have been focused on three main areas: ultrafine-grained aluminum alloys produced by severe plastic deformation; aluminum matrix composites; and aluminum-based composite materials (including material combinations such as magnesium/aluminum or steel/aluminum and the corresponding joining and forming technologies). The framework of SFB 692 has served as a base for numerous scientific collaborations between scientists in the fields of materials scientists, design engineering, forming technology, production engineering, mechanics, and even economics—in Chemnitz, and with many well-established international experts throughout the world.

While research on these topics will certainly continue in Chemnitz during the coming years, funding of our SFB 692 has run out as we have reached the funding program's regular limit. The German Research Foundation encourages a concerted effort of joint publications to summarize the scientific progress achieved in Collaborative Research Centers. This is why the governing board of SFB 692 decided in mid-2017 to organize and publish a Special Issue in the well-established open access journal, Metals, with a focus on recent results on high-strength aluminum-based light-weight materials, and with the aim of also providing a broad overview of research activities in SFB 692. As the Speaker of SFB 692, I have been pleased to see that a substantial number of manuscripts have been submitted by our researchers (many of them joint work from several projects), and almost all of them passed quickly through the external, high-quality peer review of the Metals journal. As Metals is an open access journal, all papers can be found online at: http://www.mdpi.com/journal/metals/special_issues/German_Collaborative_Research_Center_692.

It is my pleasure to present these scientific papers from this Special Issue as a printed version in this book.

Martin F.-X. Wagner

Special Issue Editor

![metals logo] *metals*   MDPI

Editorial

Light-Weight Aluminum-Based Alloys—From Fundamental Science to Engineering Applications

Martin Franz-Xaver Wagner

Institute of Materials Science and Engineering, Chemnitz University of Technology, Erfenschlager Str. 73, 09125 Chemnitz, Germany; martin.wagner@mb.tu-chemnitz.de; Tel.: +49-371-531-36153

Received: 6 April 2018; Accepted: 10 April 2018; Published: 11 April 2018

Keywords: aluminum alloys; aluminum matrix composites; aluminum-magnesium composites; severe plastic deformation; ultrafine-grained materials; equal-channel angular pressing; forming technology; materials science; surface engineering; mechanics and modeling

---

1. Introduction and Scope

Academia and industry alike are faced with an ever-growing demand for energy-efficiency and reduced mass. Aluminum-based light-weight materials offer great potential for novel engineering applications, particularly when they are optimized to exhibit high strength and yet provide sufficient reliability. The last decade has thus seen substantial activity in the research fields of high-strength aluminum alloys and aluminum-based composite materials. For twelve years, backed by substantial funding of the German Research Foundation (Deutsche Forschungsgemeinschaft, DFG, Bonn, Germany), scientists of the Collaborative Research Center "High-strength aluminum-based light-weight materials for safety components" (SFB 692) at TU Chemnitz, Germany, have contributed to this research area. Our research efforts have been focused on three main areas: ultrafine-grained (UFG) aluminum alloys produced by severe plastic deformation; aluminum matrix composites (AMCs); and aluminum-based composite materials (including material combinations like magnesium/aluminum or steel/aluminum and the corresponding joining and forming technologies). The framework of SFB 692 has served as a basis for numerous scientific collaborations between scientists from the fields of materials science, design engineering, forming technology, production engineering, mechanics, and even economics—in Chemnitz, and with many well-established experts throughout the world. This Special Issue, comprising 15 scientific papers, represents a joint effort to provide a broad overview of our research activities in the field of aluminum-based light-weight materials, and to bring some recent and relevant results to the attention of a wider international audience.

2. Contributions

A key topic studied in the Research Center is the science and engineering of ultrafine-grained materials produced by severe plastic deformation (SPD), where large amounts of plastic deformation at room temperature lead to a gradual formation of very fine grains via the stages of accumulation of dislocations in cell walls and the formation of small and large angle grain boundaries [1]. Over the years, advances in the field of SPD processing [2,3] have strongly influenced the scientific approach of our Chemnitz consortium. Three main SPD techniques are well established by now—accumulative roll bonding (for the production of UFG sheet materials), high pressure torsion (for the generation of small sample volumes with grain sizes down to a couple of nanometers or even into the amorphous state), and equal-channel angular pressing (ECAP). The latter method, pioneered in the early 1980s in Russia [4,5], has become the flagship process of Chemnitz's materials engineers. Even today, ECAP represents the only viable approach that provides relatively large volumes of bulk UFG materials—an important requirement for industrial applications that can then exploit the high strength of these

UFG structural materials. We have put much effort into up-scaling of our ECAP dies to industrially relevant scales [6,7]—in fact, with a cross-section of 50 × 50 mm², our researchers have access to an experimental setup that is one of the largest ECAP dies in the world.

Several papers in this Special Issue deal with ECAP and related SPD methods. Horn et al. [8] address heterogeneous shear deformation during ECAP. While nominally, the simple shear occurring in the ECAP die's shear zone is expected to be homogeneous, the formation of shear bands adjacent to less deformed matrix bands in a regular pattern is an interesting experimental phenomenon that raises questions on the (macro-scale) homogeneity of conventional ECAP billets. The paper summarizes a careful modeling approach that leads to a close agreement with experimental findings. Moreover, the authors' interpretation of the deformation mode in the light of bifurcation analysis may well lead to a more general description of shear instabilities during ECAP. The paper by Fritsch et al. [9] deals with low-temperature ECAP—a technique that makes it possible to process an AA7075 alloy that is not sufficiently ductile to deform by ECAP at room temperature. The paper's main focus is on the positive effects of suitable heat treatments *prior to* ECAP processing, highlighting the effect of precipitates on grain refinement during subsequent, low-temperature deformation. Berndt et al. [10] move beyond the conventional understanding of SPD processing as a simple deformation that maintains the macroscopic geometry of a billet. They report on low-temperature extrusion of an AA6060 alloy, exploring the idea of producing fine-grained or even UFG microstructures while generating billet shapes that can, in principle, be more complex than conventional ECAP billets with square or circular cross-sections. While a low-temperature extrusion process leads to microstructural gradients that can be partially homogenized by subsequent aging heat treatments, gradation extrusion, as described in the paper by Frint et al. [11], represents a forming technique that explicitly aims at creating distinct gradients. The method, developed at the Chemnitz Fraunhofer Institute for Machine Tools and Forming Technology (IWU), limits SPD to the surface regions, and hence produces billets that are fine-grained at the surface, but coarse-grained (and hence more ductile) in the center.

Magnesium alloys exhibit an even lower density than aluminum alloys, but are prone to corrosion. An obvious solution is to combine magnesium and aluminum alloys in a component, for instance, by cladding the aluminum part around a magnesium core. This approach is discussed in detail by Förster et al. [12], who performed both co-extrusion and die-forging experiments in combination with detailed numerical analysis to fully model the entire processing chain. Intermetallic phases are likely to form when different alloys are joined at elevated temperatures to produce a macroscopic component, and the growth and fragmentation of such phases play an important role during forming of the aluminum/magnesium compounds studied in our Research Center. Kirbach et al. [13] investigate the properties of the corresponding boundary layers from a fracture mechanics perspective. When dealing with UFG microstructures, thermal stability becomes an important factor, and high temperatures often cannot be used for joining processes. Consequently, Habisch et al. [14] study the potential of diffusion-bonding of aluminum alloys and magnesium with different interlayer materials. Scherzer et al. [15] report on a different kind of component; they present dedicated Finite Element simulations of the Presta joining process for assembled camshafts. Joining steel and aluminum in such a component is challenging, for instance, because of thermal expansion. The careful simulations presented here allow for a detailed analysis of all forming steps that reproduce even subtle features of the real life process chain.

Using particles (like SiC) as reinforcements represents an alternative approach for increasing the strength of aluminum alloys in AMCs. Introducing such particles, of course, leads to all sorts of novel phenomena and materials engineering challenges. In this Special Issue, Härtel et al. [16] analyze the mechanical behavior of an SiC-reinforced AA2017 alloy at different strain rates and temperatures. They document that pronounced localized deformation is related to the growth of Portevin-Le Châtelier bands. Siebeck et al. [17] consider the high-temperature deformation of different AMCs, with a specific focus on the effect of mechanical alloying of small additions of boron on creep resistance. Completing the multifaceted investigations on the thermo-mechanical behavior of bulk

AMCs, Winter et al. [18] discuss the effects of different sizes and volume fractions of SiC reinforcements, and of testing temperature, on the high cycle fatigue behavior of a reinforced AA2124 alloy.

In many engineering materials, fatigue life can be massively increased by suitable modifications of the surface. The paper by Nestler and Schubert [19] documents how roller burnishing of an AMC can produce high-quality surfaces and also introduce compressive residual stresses, which can be particularly beneficial for improving the fatigue behavior. Surface properties also come into play when wear and corrosion dominate a component's life span. Sieber et al. [20] show that hard anodizing in sulphuric acid is an energy-efficient process that provides good abrasion resistance of conversion coatings. In ever-more complex combinations of materials in machine parts and components, the need may arise to combine AMCs with either aluminum alloys or with steel. Grund et al. [21] analyze how arc brazing can be used as a joining method at relatively low temperatures in such cases. As materials scientists and engineers, we are often intrigued when new processing techniques make it possible to produce novel materials with interesting microstructures. This may be particularly true for composite materials, where the sheer number of possible combinations and variations of matrix materials and reinforcements is staggering. Yet, considering properties in relation to costs, many of these materials will never make it into real-world applications. The detailed economic analysis by Schmidt et al. [22], combining technology, user and market analyses, is all the more encouraging: it emphasizes the economic potential of AMCs, particularly related to their high strength and weight reduction.

3. Conclusions and Outlook

The Collaborative Research Center SFB 692 has, at this point, reached the funding program's regular limit—so where will we go from here? Research on many of the topics highlighted in this Special Issue will certainly continue in Chemnitz during the coming years. In the field of SPD methods, we have demonstrated that reliable, reproducible application of ECAP still poses many challenges, from materials engineering to physical metallurgy. Likewise, the development of new AMCs with improved properties is far from finished, and light-weight components made of multiple materials, for instance in hybrid structures, are currently a hot topic both for industry and fundamental science. Making materials lighter, yet strong and reliable, will certainly remain an attractive goal for scientific endeavors for many years to come. As former Speaker of the Collaborative Research Center SFB 692, I do hope that this Special Issue promotes the ideas developed in Chemnitz and in our extended scientific network, and that many readers find our papers of sufficient interest to stimulate ideas of their own.

Conflicts of Interest: The author declares no conflict of interest.

References

1. Hughes, D.A.; Hansen, N. High angle boundaries formed by grain subdivision mechanisms. *Acta Mater.* **1997**, *45*, 3871–3886. [CrossRef]
2. Valiev, R. Nanostructuring of metals by severe plastic deformation for advanced properties. *Nat. Mater.* **2004**, *3*, 511–516. [CrossRef] [PubMed]
3. Estrin, Y.; Vinogradov, A. Extreme grain refinement by severe plastic deformation: A wealth of challenging science. *Acta Mater.* **2013**, *61*, 782–817. [CrossRef]
4. Segal, V.M. Materials processing by simple shear. *Mater. Sci. Eng. A* **1995**, *197*, 157–164. [CrossRef]
5. Segal, V.M. Equal channel angular extrusion: From macromechanics to structure formation. *Mater. Sci. Eng. A* **1999**, *271*, 322–333. [CrossRef]
6. Frint, P.; Hockauf, M.; Halle, T.; Strehl, G.; Lampke, T.; Wagner, M.F.-X. *Microstructural Features and Mechanical Properties after Industrial Scale ECAP of an Al-6060 Alloy*; Trans Tech Publications: Zürich, Switzerland, 2011; Materials Science Forum Volumes 667–669; ISBN 9783037850077.
7. Frint, S.; Hockauf, M.; Frint, P.; Wagner, M.F.-X. Scaling up Segal's principle of Equal-Channel Angular Pressing. *Mater. Des.* **2016**, *97*. [CrossRef]

8. Horn, T.D.; Silbermann, C.B.; Frint, P.; Wagner, M.F.-X.; Ihlemann, J. Strain localization during equal-channel angular pressing analyzed by finite element simulations. *Metals* **2018**, *8*, 55. [CrossRef]
9. Fritsch, S.; Wagner, M.F.-X. On the effect of natural aging prior to low temperature ECAP of a high-strength aluminum alloy. *Metals* **2018**, *8*, 63. [CrossRef]
10. Berndt, N.; Frint, P.; Wagner, M.F.-X. Influence of extrusion temperature on the aging behavior and mechanical properties of an AA6060 aluminum alloy. *Metals* **2018**, *8*, 51. [CrossRef]
11. Frint, P.; Härtel, M.; Selbmann, R.; Dietrich, D.; Bergmann, M.; Lampke, T.; Landgrebe, D.; Wagner, M.F.X. Microstructural evolution during severe plastic deformation by gradation extrusion. *Metals* **2018**, *8*, 96. [CrossRef]
12. Förster, W.; Binotsch, C.; Awiszus, B. Process Chain for the Production of a Bimetal Component from Mg with a Complete Al Cladding. *Metals* **2018**, *8*, 97. [CrossRef]
13. Kirbach, C.; Stockmann, M.; Ihlemann, J. A Fragmentation Criterion for the Interface of a Hydrostatic Extruded Al-Mg-Compound. *Metals* **2018**, *8*, 157. [CrossRef]
14. Habisch, S.; Böhme, M.; Peter, S.; Grund, T.; Mayr, P. The Effect of Interlayer Materials on the Joint Properties of Diffusion-Bonded Aluminium and Magnesium. *Metals* **2018**, *8*, 138. [CrossRef]
15. Scherzer, R.; Fritsch, S.; Landgraf, R.; Ihlemann, J.; Wagner, M.F.-X. Finite element simulation of the presta joining process for assembled camshafts: Application to aluminum shafts. *Metals* **2018**, *8*, 128. [CrossRef]
16. Härtel, M.; Illgen, C.; Frint, P.; Wagner, M.F.-X. On the PLC effect in a particle reinforced AA2017 alloy. *Metals* **2018**, *8*, 88. [CrossRef]
17. Siebeck, S.; Roder, K.; Wagner, G.; Nestler, D. Influence of boron on the creep behavior and the microstructure of particle reinforced aluminum matrix composites. *Metals* **2018**, *8*, 110. [CrossRef]
18. Winter, L.; Hockauf, K.; Lampke, T. Temperature and Particle Size Influence on the High Cycle Fatigue Behavior of the SiC Reinforced 2124 Aluminum Alloy. *Metals* **2018**, *8*, 43. [CrossRef]
19. Nestler, A.; Schubert, A. Roller Burnishing of Particle Reinforced Aluminium Matrix Composites. *Metals* **2018**, *8*, 95. [CrossRef]
20. Sieber, M.; Morgenstern, R.; Scharf, I.; Lampke, T. Effect of nitric and oxalic acid addition on hard anodizing of $AlCu_4Mg_1$ in sulphuric acid. *Metals* **2018**, *8*, 139. [CrossRef]
21. Grund, T.; Gester, A.; Wagner, G.; Habisch, S.; Mayr, P. Arc Brazing of Aluminium, Aluminium Matrix Composites and Stainless Steel in Dissimilar Joints. *Metals* **2018**, *8*, 166. [CrossRef]
22. Schmidt, A.; Siebeck, S.; Götze, U.; Wagner, G.; Nestler, D. Particle-Reinforced Aluminum Matrix Composites (AMCs)—Selected Results of an Integrated Technology, User, and Market Analysis and Forecast. *Metals* **2018**, *8*, 143. [CrossRef]

metals

MDPI

Article

Influence of Extrusion Temperature on the Aging Behavior and Mechanical Properties of an AA6060 Aluminum Alloy

Nadja Berndt, Philipp Frint * and Martin F.-X. Wagner

Institute of Materials Science and Engineering, Chemnitz University of Technology, Erfenschlager Str. 73, 09125 Chemnitz, Germany; nadja.berndt@mb.tu-chemnitz.de (N.B.); martin.wagner@mb.tu-chemnitz.de (M.F.-X.W.)
* Correspondence: philipp.frint@mb.tu-chemnitz.de; Tel.: +49-371-531-38218

Received: 12 December 2017; Accepted: 9 January 2018; Published: 12 January 2018

Abstract: Processing of AA6060 aluminum alloys for semi-products usually includes hot extrusion with subsequent artificial aging for several hours. Processing below the recrystallization temperature allows for an increased strength at a significantly reduced annealing time by combining strain hardening and precipitation hardening. In this study, we investigate the potential of cold and warm extrusion as alternative processing routes for high strength aluminum semi-products. Cast billets of the age hardening aluminum alloy AA6060 were solution annealed and then extruded at room temperature, 120 or 170 °C, followed by an aging treatment. Electron microscopy and mechanical testing were performed on the as-extruded as well as the annealed materials to characterize the resulting microstructural features and mechanical properties. All of the extruded profiles exhibit similar, strongly graded microstructures. The strain gradients and the varying extrusion temperatures lead to different stages of dynamic precipitation in the as-extruded materials, which significantly alter the subsequent aging behavior and mechanical properties. The experimental results demonstrate that extrusion below recrystallization temperature allows for high strength at a massively reduced aging time due to dynamic precipitation and/or accelerated precipitation kinetics. The highest strength and ductility were achieved by extrusion at 120 °C and subsequent short-time aging.

Keywords: aluminum alloy; cold extrusion; warm extrusion; severe plastic deformation (SPD); dynamic aging; precipitation hardening

---

1. Introduction

Aluminum alloys of the 6xxx series, which are prominent examples for many structural applications in automotive and aviation industry, are usually processed by hot extrusion. Their mechanical properties are then adjusted by a subsequent heat treatment since the alloying elements Mg and Si enable an effective precipitation hardening the low flow stress at elevated temperatures allows for high extrusion ratios (ratio of cross sectional area of the billet versus the extrudate) as well as pressing speeds. For processing of the AA6060 aluminum alloy, which has a very high formability, and therefore is one of the most commonly used aluminum alloys, the homogenized billets are typically extruded at temperatures between 400 and 500 °C, followed by water or air quenching [1]. Enhancing the strength of the extruded semi-products by artificial aging at 160 to 180 °C takes up several hours. One effective, alternative processing strategy that considerably reduces aging times and additionally increases strength is cold deformation of the as-solutionized material. The high densities of dislocations and vacancies produced by large plastic strains ensure effective strain hardening; they also increase the diffusion rates of solute atoms, and can therefore significantly accelerate precipitation kinetics [2,3]. In case of severe plastic deformation (SPD), e.g., by equal-channel angular pressing (see e.g., [4–6]) or high-pressure torsion, a rearrangement of dislocations into cell walls and subsequently into small and large angle grain boundaries results in the formation of fairly homogeneous ultrafine-grained (UFG) microstructures [7]. At elevated temperatures, partial or complete dynamic precipitation may also occur during SPD [8–11]. The strengthening effect of subsequent artificial aging therefore strongly depends on the amount of applied (equivalent) strain, as well as on the aging temperature. The strengthening effects is, however, often limited: age hardening can be suppressed or even replaced by age softening [12–14].

In our previous study [15], we investigated an unconventional processing technique for the AA6060 aluminum alloy that aimed at combining the benefits of SPD processing (high strength, short aging time) and extrusion (change of shape) by extruding below recrystallization temperature. We characterized two ways of processing that significantly increase hardness and strength when compared to the conventional T6 condition—sequential and simultaneous extrusion and aging, respectively. While the first—room-temperature extrusion and subsequent aging—results in a more homogenous distribution of strength and a slightly higher ductility, the second—extrusion at aging temperature (170 °C)—benefits from lower pressing forces and advanced dynamic precipitation.

As none of these strategies fully exploit the potential of combined strain and precipitation hardening, the present study is focused on the optimization of mechanical properties by extrusion at 120 °C, followed by a suitable heat treatment. We compare the microstructural features and aging behavior of the extruded material with the results from our previous study and we show how our optimized processing route allows for the fabrication of rod-shaped semi-products with a homogeneously distributed maximum hardness as well as strength and a reasonable ductility.

2. Materials and Methods

The material used in this study is the precipitation hardenable aluminum alloy AA6060, with a chemical composition of 0.5Si-0.5Mg-0.2Cu-0.2Fe-balance Al (wt %). Billets of the continuous cast material were solid-solution annealed for 3 h at 530 °C, and then water-quenched to room temperature. Backward-extrusion was carried out in horizontal extrusion press with a maximum press capacity of 8 MN. The diameters of the cast billets and of the containers were 107 and 110 mm, respectively. A flat die with an inner diameter of 45 mm was used, resulting in an extrusion ratio (ratio of billet cross-section area vs. extrudate cross-section area) of about 6.

Three different processing temperatures were selected for extrusion. On the one hand, extrusion was performed at room temperature (RT), which results in maximum strain hardening of the aluminum alloy. On the other hand, extrusion was performed at elevated temperatures of 120 (intermediate temperature—IT) and 170 °C (which corresponds to the conventional aging temperature of this alloy—AT) in order to reduce pressing forces and to promote dynamic precipitation. Because it directly combines two essential processing steps, extrusion at these temperatures presents an economically promising approach. To prevent an excessive temperature increase related to friction and quasi-adiabatic heating, extrusion was carried out at a low ram speed of 18 mm/min. In addition, Bechem Beruforge 150D, a lubricant that is well suited for cold forming of aluminum [16], was applied to the contact areas between the billets and the die. For AT-extrusion, the billets were heated to 170 °C in an induction furnace, while the die was heated in a convection furnace prior to processing. The container was held at a temperature of 150 °C during the entire pressing operation. For IT-extrusion, all processing temperatures (billet, die, and container) were heated to 120 °C. The IT- and AT-extrudates were spray-cooled at the runout of the press.

The aging behavior of the extruded material was studied after artificial aging (performed in a convection furnace at 170 °C, with annealing times ranging from 1 to 6000 min) using an automatic hardness tester (KB250BVRZ, KB Prüftechnik GmbH, Hochdorf-Assenheim, Germany). Vickers hardness (HV1) was measured on the longitudinal planes of the extrudates (i.e., planes containing the extrusion direction and a radial direction). Flat specimens for tensile testing were cut along the extrusion direction using wire electric discharge machining (see Figure 1). Quasi-static tensile testing was performed at room temperature with an initial strain rate of 10^{-3} s^{-1} using a Zwick/Roell Z020 universal testing machine with a 20 kN load cell. Strains were measured using a digital image correlation system (GOM GmbH, Braunschweig, Germany).

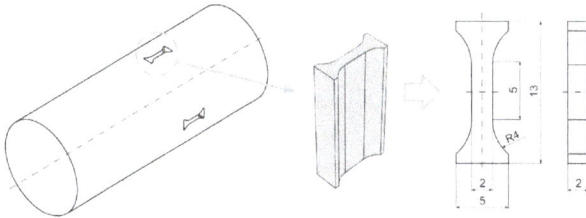

Figure 1. Schematic representation of sample extraction for tensile testing. Flat miniature tensile specimens were cut from the extruded material using wire electric discharge machining.

Samples for microstructural analysis by scanning electron microscopy (SEM) and transmission electron microscopy (TEM) were extracted from the transverse planes at the centers and in the peripheral areas of the extrudates. For electron back-scatter diffraction (EBSD) measurements, the samples were mechanically prepared by a standard grinding and polishing procedure with an additional vibratory polishing step for one hour in aqueous colloidal SiO$_2$ solution. The EBSD patterns were collected with an EDAX detector using a Zeiss Neon 40 field emission scanning electron microscope at an acceleration voltage of 10 kV and with an aperture of 120 μm. Post-processing included a slight clean-up of the raw data (neighbor confidence index correlation). The TEM samples were mechanically thinned to approximately 80 μm followed by twin jet electro-polishing (Tenupol-5, electrolyte: A7, −32 °C). TEM was carried out using a Hitachi H8100 TEM with an acceleration voltage of 200 kV.

3. Results and Discussion

3.1. Microstructural and Mechanical Characterization of the as-Extruded Material

As discussed in detail in our previous study [15], cold and warm extrusion of an AA6060 aluminum alloy results in relatively similar, heterogeneous microstructures that exhibit two characteristic regions: The majority of the extrudate's cross sectional area consists of a microstructure containing both coarse and fine grains that are elongated parallel to the direction of extrusion. Towards the surface, a distinct layer with a width of about 2 mm is observed. This layer exhibits an (ultra)fine-grained pancake-microstructure, i.e., the grains are elongated parallel to the direction of extrusion, as well as parallel to the circumference of the extrudate [15].

The microstructures of the center areas and surface layers of the as-extruded conditions are shown in Figure 2. The (large angle) grain boundary maps (obtained by EBSD) show a similar microstructure for the center areas (Figure 2a–c), where areas of large grains can be found next to areas containing considerably smaller grains. In the surface layers, a similar bimodal-like microstructure can be observed. However, the overall grain size is much smaller, i.e., areas of fine grains exist next to UFG areas (Figure 2d–f). The fraction of UFG areas is the highest after RT-extrusion with about 50% of the area shown in the grain boundary map (Figure 2d). In contrast, the UFG areas of the IT- and AT-extruded materials correspond to only about 20% of the scanned area (Figure 2e,f). This indicates an earlier stage of the grain refinement process after warm extrusion, which is most likely caused by the more pronounced recovery during extrusion at elevated temperatures.

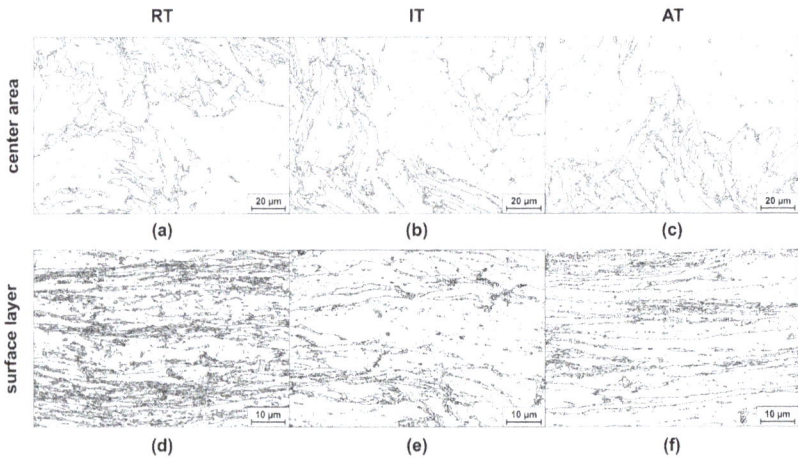

Figure 2. Microstructure of the center areas (**a**–**c**) and surface layers (**d**–**f**) of the extruded materials. The grain boundary maps were obtained by electron back-scatter diffraction (EBSD) (step size of 500 nm for the center areas and of 100 nm for the surface layers) and show the distribution of large angle grain boundaries with a minimum misorientation angle of 15°.

Further differences regarding the microstructure after cold and warm extrusion are shown in Figure 3. The TEM bright-field micrographs were taken from the center areas and surface regions of the three extrudates and illustrate the effect of dynamic precipitation during deformation at elevated temperatures. While no precipitates could be found in the center area or the surface layer of the RT-extruded material (Figure 3a,d), precipitates of varying size were found in both of the regions after IT- (Figure 3b,e) and AT-extrusion (Figure 3c,f). For extrusion at both 120 and 170 °C, the center areas exhibited small precipitates in the grain interiors (areas marked by dashed white ellipses) and slightly larger ones in the proximity of grain boundaries (see black ellipse in Figure 3b). Please note that the size of the coherent β″-precipitates is rather low (needle-shaped with several ten nanometers in length and less than ten nanometers in diameter), which makes them difficult to recognize in the TEM micrographs, In the surface layers precipitation often took place in close proximity to the grain boundaries (dashed black ellipses in Figure 3e) and overall larger precipitates—when compared to the center areas—were observed. For AT-extrusion, cuboid-shaped precipitates with a diameter of about 80 nm were observed in the surface layer (Figure 3f, dashed black circle), which strongly indicates the beginning of over-aging in this region.

In order to characterize the mechanical properties of the extruded materials, Vickers hardness was measured as a function of distance from the extrudates' surfaces (Figure 4). For the RT-extruded material, the lowest hardness occurs in the center area and then increases towards the periphery, corresponding to an increasing amount of plastic deformation. Near the surface layer, starting at a depth of approximately 2–2.5 mm, a sharp increase of hardness is observed; this increase likely results from an additional grain boundary strengthening effect due to the (ultra)fine-grained microstructure. A similar surface layer hardness of about 103 HV was measured for all as-extruded conditions. In contrast to the hardness profile of the RT-extruded material, however, the decrease of hardness towards the center is not as pronounced after IT-extrusion, and the AT-extruded material actually exhibits a slightly increasing hardness from the surface towards the center of the extrudate. Therefore, despite the UFG microstructure that was also observed in its surface layer, a higher hardness could be obtained in the coarser grained center area. This effect is most probably related to the different stages of precipitation in the core when compared to the surface of the AT-extruded material, as also indicated by the microstructural features documented in Figure 3c,f.

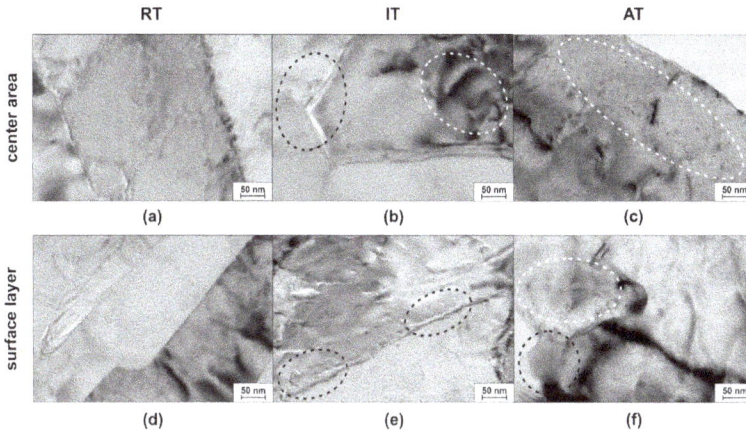

Figure 3. Transmission electron microscopy (TEM) bright-field images taken from the center areas (**a–c**) and surface layers (**d–f**) of the extruded material. While room temperature (RT)-extrusion does not initiate dynamic precipitation, precipitates with diameters ranging from a few nanometers up to 80 nm were found both in the IT- and AT-extruded materials (areas containing precipitates are highlighted by white and black ellipses).

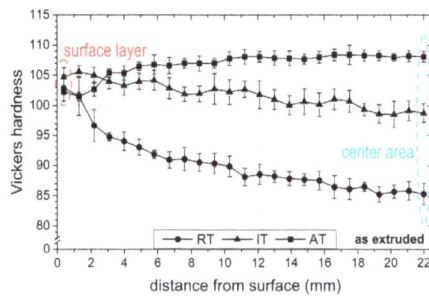

Figure 4. Vickers hardness as a function of distance from surface for the as-extruded materials after RT- (circles), IT- (120 °C, triangles) and AT-extrusion (170 °C, squares). Dashed red and blue ellipses highlight hardness values that represent the surface layers and the center areas, respectively.

3.2. Aging Behavior of Cold and Hot Extruded Materials

To study the aging behavior of the surface layers and the center areas, Vickers hardness was measured after aging for different times at 170 °C (Figure 5). The first data points (at an aging time of 0.01 min) represent the first and last points of the hardness distribution curves of the as-extruded conditions, as marked by the dashed red and blue ellipses in Figure 4. Although the initial core hardness (blue curves) differs strongly for the investigated extrusion temperatures, a similar maximum hardness of about 109 to 115 HV can be achieved at aging times that range from 40 to 90 min. In contrast, the three extrudates exhibit a similar initial surface hardness (red curves), which then develops diversely during aging: While it is possible to increase the hardness of the surface layer after IT-extrusion up to 115 HV by artificial aging for about 40 min, age softening occurs after RT- and AT-extrusion. In the case of AT-extrusion, the softening effect sets in with the beginning of the additional annealing, which corresponds to the beginning of over-aging, as already discussed on the basis of the TEM micrographs (Figure 3). For the RT-extruded material, hardness remains constant for about 100 min before significant softening sets in. As discussed in [15], this probably results from a dynamic equilibrium of the competing processes of precipitation hardening and softening by temperature-driven recovery. As it is likely to have already recovered during warm extrusion, and because it does not exhibit unusually large precipitates like the AT-extruded material, the surface layer of the IT-extruded material maintains its potential for precipitation hardening.

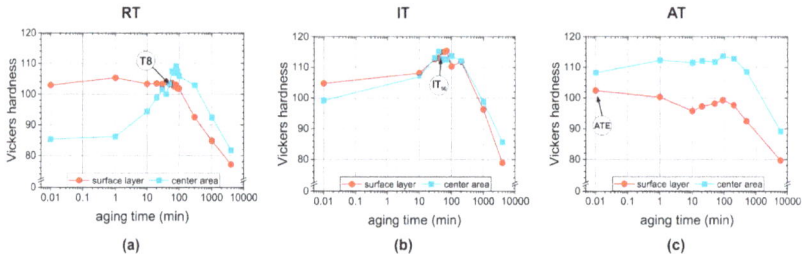

Figure 5. Vickers hardness plots of the surface layers and the center areas of the extruded materials as a function of aging time at 170 °C for (**a**) RT-extrusion; (**b**) IT-extrusion (120 °C); and, (**c**) AT-extrusion (170 °C). Material conditions with the highest homogenous hardness ("optimized" material conditions) are marked for all the extruded materials (labels in circles).

3.3. Microstructural and Mechanical Characterization of Optimized Material Conditions

On the basis of the hardness data presented in Figure 5, one condition with the highest homogeneous hardness was chosen as "optimized" material for each extrusion temperature: RT-extruded and peak-aged (50 min at 170 °C; labeled T8 in Figure 5), IT-extruded and aged for 50 min (labeled IT_{50}) and AT-extruded without subsequent aging (as-extruded; labeled ATE). Additional TEM investigations were performed to relate microstructural changes during aging with the documented changes in terms of mechanical properties. TEM analysis of the optimized conditions T8 and IT_{50} (Figure 6) showed that artificial aging for 50 min results in finely dispersed precipitates that are responsible for the increased hardness values. Precipitation occurred both in the center areas and in the surface layers, therefore contributing to an improved homogeneity of hardness values, particularly in the RT-extruded material condition.

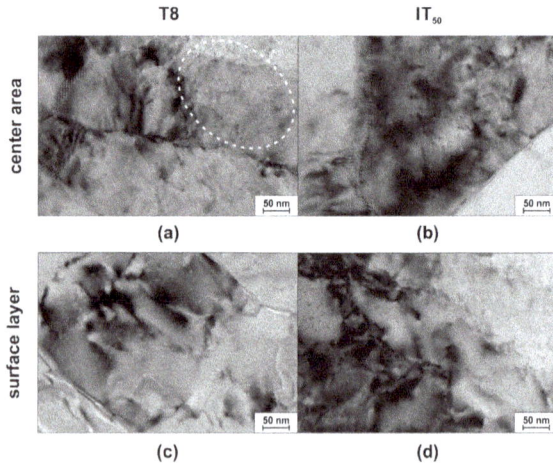

Figure 6. TEM bright-field images taken from the center areas (**a,b**) and surface layers (**c,d**) of the extruded and artificially aged material. Both conditions T8 and IT_{50} exhibit very fine precipitates in the center areas as well as in the surface layers.

The tensile properties were characterized in great detail by examining tensile test data as a function of distance from the extrudate surface. The results are summarized in terms of yield strength (YS), ultimate tensile strength (UTS), uniform elongation (UE), and elongation to failure (ETF) in Figure 7. For the ATE material condition, the evolution of YS is similar to the corresponding hardness curve, with the lowest value near the surface (308 MPa) and increasing values towards the center (>350 MPa). In contrast, the YS of both aged conditions T8 and IT_{50} can be considered constant over the cross section despite the differences in microstructure. As shown in Figure 7b, the UTS of all the investigated conditions is similarly distributed over the cross section with values increased by about 15 to 20 MPa when compared to the respective YS. The highest UTS of 380 MPa is measured for the IT-extruded and artificially aged material. When compared to the peak-aged (2000 min at 170 °C) cast material with an UTS of 289 MPa, this represents an increase by about 31%. Regarding UE, all conditions exhibit a similar distribution with lowest values near the surface and increased ductility in the center area. The lower ductility in the surface layer is likely to result from the maximum amount of cold work in this region, which leads to a low hardening rate and to early necking of the specimens during tensile testing. The overall somewhat better uniform elongation of the T8 and IT_{50} conditions is assumed to result from the additional heat treatment that allowed for a reduction of lattice defects by recovery mechanisms.

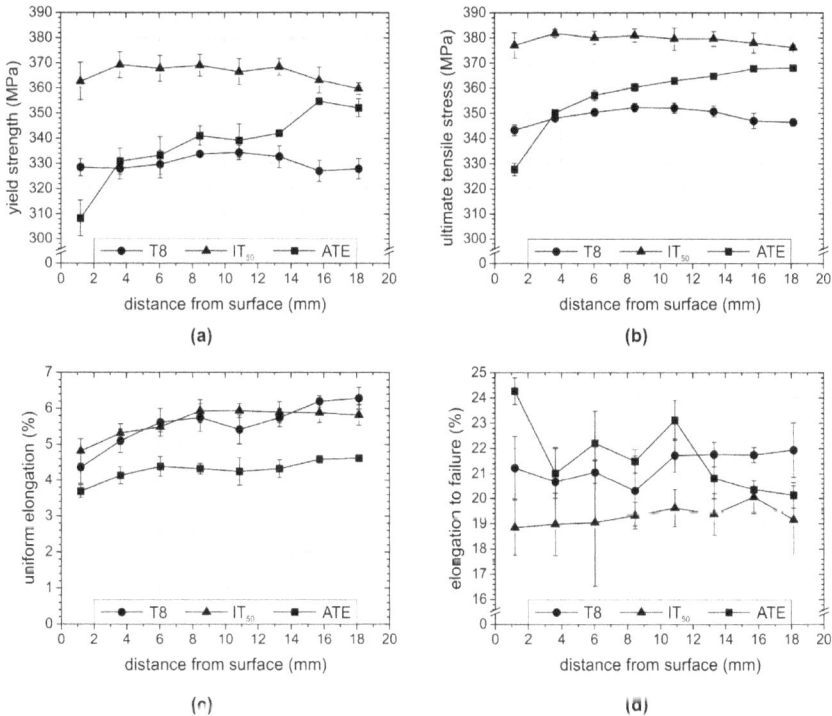

Figure 7. Overview of the tensile properties of the optimized conditions T8 (RT-extrusion and peak aged, circles), IT_{50} (IT-extrusion and aged for 50 min, triangles) and ATE (AT-extrusion, as-extruded, squares) as a function of distance from the surface. (**a**) Yield strength; (**b**) ultimate tensile strength; (**c**) uniform elongation; and, (**d**) elongation to failure.

4. Summary and Conclusions

We have studied the effect of extrusion temperature on the aging behavior and the resulting mechanical properties of cylindrical rods that were extruded from cast billets of an AA6060 aluminum alloy at room temperature (RT), at 120 °C (IT), and at 170 °C (AT), respectively. As discussed in our previous study [15], cold and warm extrusion of this alloy results in heterogeneous microstructures with rather coarse elongated grains in the center areas, and a partially ultrafine-grained pancake microstructure in the surface layers, of the extrudates. Due to the different processing temperatures and the gradient of plastic straining between core and surface of the extrudates, various stages of the precipitation sequence were observed in the as-extruded material. We found that the extent of strain hardening and dynamic precipitation significantly alter the effects of a subsequent artificial aging. While AT-extrusion fully exploits the effects of dynamic aging—resulting, e.g., in the lowest processing time—it leads to a high, but heterogeneous strength with the lowest ductility (uniform elongation). Processing at RT with a subsequent short-time aging for 50 min at 170 °C offers an evenly distributed strength at a similar level and a higher ductility. When considering these parameters, the best mechanical properties were obtained by extrusion at IT and short-time aging (50 min), which offers an increase of strength by about 31% at a reduction of aging time by 97.5%.

Acknowledgments: The authors gratefully acknowledge funding by the German Research Foundation (Deutsche Forschungsgemeinschaft, DFG) in the framework of the Collaborative Research Center SFB 692 (projects T4 and Z2). We also thank Marcus Böhme and Anne Schulze for support with the electron microscopy investigations.

Author Contributions: Nadja Berndt and Philipp Frint conceived and designed the experiments. Nadja Berndt performed and analyzed the experiments, prepared the figures and drafted the manuscript. Philipp Frint and Martin F.-X. Wagner conceived of the study, discussed the results and analysis and helped writing the manuscript.

Conflicts of Interest: The authors declare no conflict of interest.

References

1. Sheppard, T. *Extrusion of Aluminium Alloys*; Kluwer Academic Publishers: Dordrecht, The Netherlands, 1999; ISBN 978-0412590702.
2. Atkinson, A.; Taylor, R.I. The diffusion of Ni in the bulk and along dislocations in NiO single crystals. *Philos. Mag. A* **1979**, *39*, 581–595. [CrossRef]
3. Staab, T.E.M.; Haaks, M.; Modrow, H. Early precipitation stages of aluminum alloys—The role of quenched-in vacancies. *Appl. Surf. Sci.* **2008**, *255*, 132–135. [CrossRef]
4. Frint, P.; Hockauf, M.; Halle, T.; Strehl, G.; Lampke, T.; Wagner, M.F.-X. Microstructural Features and Mechanical Properties after Industrial Scale ECAP of an Al-6060 Alloy. *Mater. Sci. Forum* **2011**, *667–669*, 1153–1158. [CrossRef]
5. Frint, P.; Hockauf, M.; Halle, T.; Wagner, M.F.-X.; Lampke, T. The role of backpressure during large scale equal-channel angular pressing. *Materwiss. Werkst.* **2012**, *43*, 668–672. [CrossRef]
6. Frint, S.; Hockauf, M.; Frint, P.; Wagner, M.F.-X. Scaling up Segal's principle of Equal-Channel Angular Pressing. *Mater. Des.* **2016**, *97*, 502–511. [CrossRef]
7. Valiev, R.Z.; Langdon, T.G. Principles of equal-channel angular pressing as a processing tool for grain refinement. *Prog. Mater. Sci.* **2006**, *51*, 881–981. [CrossRef]
8. Sha, G.; Wang, Y.B.; Liao, X.Z.; Duan, Z.C.; Ringer, S.P.; Langdon, T.G. Influence of equal-channel angular pressing on precipitation in an Al–Zn–Mg–Cu alloy. *Acta Mater.* **2009**, *57*, 3123–3132. [CrossRef]
9. Roven, H.J.; Liu, M.; Werenskiold, J.C. Dynamic precipitation during severe plastic deformation of an Al–Mg–Si aluminium alloy. *Mater. Sci. Eng. A* **2008**, *483–484*, 54–58. [CrossRef]
10. Gubicza, J.; Schiller, I.; Chinh, N.Q.; Illy, J.; Horita, Z.; Langdon, T.G. The effect of severe plastic deformation on precipitation in supersaturated Al–Zn–Mg alloys. *Mater. Sci. Eng. A* **2007**, *460–461*, 77–85. [CrossRef]
11. Cai, M.; Field, D.P.; Lorimer, G.W. A systematic comparison of static and dynamic ageing of two Al–Mg–Si alloys. *Mater. Sci. Eng. A* **2004**, *373*, 65–71. [CrossRef]
12. Hirosawa, S.; Hamaoka, T.; Horita, Z.; Lee, S.; Matsuda, K.; Terada, D. Methods for designing concurrently strengthened severely deformed age-hardenable aluminum alloys by ultrafine-grained and precipitation hardenings. *Metall. Mater. Trans. A* **2013**, *44*, 3921–3933. [CrossRef]

13. Hockauf, M.; Meyer, L.W.; Zillmann, B.; Hietschold, M.; Schulze, S.; Krüger, L. Simultaneous improvement of strength and ductility of Al–Mg–Si alloys by combining equal-channel angular extrusion with subsequent high-temperature short-time aging. *Mater. Sci. Eng. A* **2009**, *503*, 167–171. [CrossRef]
14. Kim, W.J.; Kim, J.K.; Park, T.Y.; Hong, S.I.; Kim, D.I.; Kim, Y.S.; Lee, J.D. Enhancement of Strength and Superplasticity in a 6061 Al Alloy Processed by Equal-Channel-Angular-Pressing. *Metall. Mater. Trans. A* **2002**, *33*, 3155–3164. [CrossRef]
15. Berndt, N.; Frint, P.; Böhme, M.; Wagner, M.F.-X. Microstructure and Mechanical Properties of an AA6060 Aluminum Alloy after Cold and Warm Extrusion. *Mater. Sci. Eng. A* **2017**, *707*, 717–724. [CrossRef]
16. Frint, P.; Wagner, M.F.-X.; Weber, S.; Seipp, S.; Frint, S.; Lampke, T. An experimental study on optimum lubrication for large-scale severe plastic deformation of aluminum-based alloys. *J. Mater. Process. Technol.* **2017**, *239*, 222–229. [CrossRef]

metals

MDPI

Article

Strain Localization during Equal-Channel Angular Pressing Analyzed by Finite Element Simulations

Tobias Daniel Horn [1,*], Christian Bert Silbermann [1], Philipp Frint [2], Martin Franz-Xaver Wagner [2] and Jörn Ihlemann [1]

1   Chair of Solid Mechanics, Chemnitz University of Technology, Reichenhainer Str. 70,
     D-09126 Chemnitz, Germany; christian.silbermann@mb.tu-chemnitz.de (C.B.S.);
     joern.ihlemann@mb.tu-chemnitz.de (J.I.)
2   Chair of Materials Science, Institute of Material Science and Engineering, Chemnitz University of
     Technology, Erfenschlager Str. 73, D-09125 Chemnitz, Germany; philipp.frint@mb.tu-chemnitz.de (P.F.);
     martin.wagner@mb.tu-chemnitz.de (M.F.-X.W.)
*   Correspondence: tobias.horn@mb.tu-chemnitz.de; Tel.: +49-371-531-31730

Received: 19 December 2017; Accepted: 8 January 2018; Published: 15 January 2018

Abstract: Equal-Channel Angular Pressing (ECAP) is a method used to introduce severe plastic deformation into a metallic billet without changing its geometry. In special cases, strain localization occurs and a pattern consisting of regions with high and low deformation (so-called shear and matrix bands) can emerge. This paper studies this phenomenon numerically adopting two-dimensional finite element simulations of one ECAP pass. The mechanical behavior of aluminum is modeled using phenomenological plasticity theory with isotropic or kinematic hardening. The effects of the two different strain hardening types are investigated numerically by systematic parameter studies: while isotropic hardening only causes minor fluctuations in the plastic strain fields, a material with high initial hardening rate and sufficient strain hardening capacity can exhibit pronounced localized deformation after ECAP. The corresponding finite element simulation results show a regular pattern of shear and matrix bands. This result is confirmed experimentally by ECAP-processing of AA6060 material in a severely cold worked condition, where microstructural analysis also reveals the formation of shear and matrix bands. Excellent agreement is found between the experimental and numerical results in terms of shear and matrix band width and length scale. The simulations provide additional insights regarding the evolution of the strain and stress states in shear and matrix bands.

Keywords: equal-channel angular pressing; ECAP; shear band; matrix band; kinematic hardening; FEM; strain localization

---

1. Introduction

Equal-Channel Angular Pressing (ECAP) is a severe plastic deformation (SPD) process developed by Segal [1,2]. During ECAP, a billet with a typically square or circular cross-section is pressed through an angled channel. Both channels (entrance and exit channel) are intersected by a shear plane, where a shear deformation is introduced into the material. The angle between these two channels (Φ) and the angle of the outer curvature (Ψ) inside the "L"-shaped channel (cf. Figure 1) control the amount of introduced (effective) strain [3].

In the most common ECAP dies ($\Phi = 90°$, $\Psi = 0°$), the introduced effective strain is approximately 1.15 after a single pass. As a consequence of the severe plastic deformation, grain refinement occurs. Grain refinement results in a higher yield strength since geometrical boundaries are effective obstacles for dislocation motion [4,5]. In addition, an extraordinarily high ductility is observed in many materials after multiple ECAP passes [6–8].

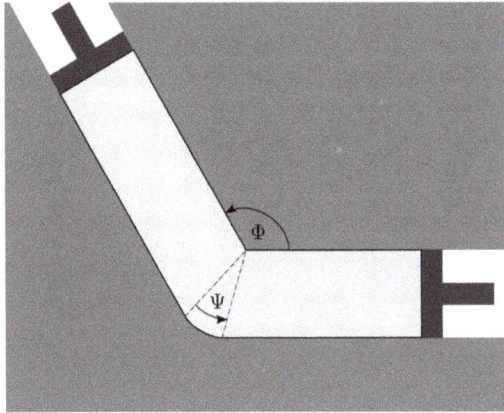

Figure 1. Sketch of an Equal-Channel Angular Pressing (ECAP) channel with the channel angle Φ and the angle Ψ defining the outer curvature.

For most metallic materials, the billet is deformed predominantly homogeneously during ECAP [8–10]. However, under certain conditions, pronounced strain localization occurs and results in heterogeneous microstructures consisting of regions with large inherent strain (shear bands) and regions with much lower strains (matrix bands) [11–15].

In addition to experimental investigations, finite element (FE) simulations have been carried out to understand the ECAP process in greater detail. For this purpose, multiple parameter studies were conducted, e.g., studying the effect of variations of the channel angle Φ [16–19], the angle of the outer curvature Ψ [18–23], the strain rate [24–26] as well as the friction between the channel and the billet [16,18,23,26,27]. However, within this huge body of scientific work, only a few papers address the simulation of strain localization [24,28,29]. In those papers, strain softening behavior had to be used in order to simulate strain localization in distinct bands. Figueiredo et al. showed that the occurrence of shear bands depends on the initial flow-softening rate and the steady-state stress [28] as well as the strain rate [24]. Furthermore, most of the aforementioned finite element studies do not contain a convergence study for their simulation model and in some cases an automatic remeshing function was used to reach a stable simulation [18,20,24,29].

The present paper shows both experimentally and numerically that strain localization during ECAP can even occur for a strain *hardening* material. To this end, explicit finite element simulations are conducted. In order to achieve reliable FE solutions, systematic convergence studies are performed. As a special feature, the billet is meshed with pre-oriented rhomboid-shaped elements such that remeshing is not necessary. The role of two types of hardening—isotropic and kinematic—is investigated numerically by varying the corresponding material parameters. It is found that only *kinematic* hardening leads to the emergence of localized deformation in a form of shear and matrix bands. The simulation results are compared thoroughly to corresponding ECAP experiments on billets made of an AA6060 alloy. Finally, the origin of the heterogeneous plastic flow is discussed both from a microstructural and a mechanical point of view, representing a starting point for further basic research on localization phenomena during SPD processing.

2. Simulation Model

Isothermal and frictionless 2D plane-strain finite element simulations of one pass of ECAP were conducted using the commercial explicit finite element code Abaqus (Version 6.14-4). For the contact between ECAP tool and billet, the Abaqus penalty algorithm is used employing a contact stiffness factor of 0.8. The geometry parameters specifying the ECAP process are shown in Figure 2: *d* is the diameter of the billet with cross-sectional area of $d \times d$ and length h. Φ is the channel angle, Ψ the angle of the outer curvature, r is the outer and R the inner edge radius. The outer edge radius is defined via the diameter d, the angle of the outer curvature and the channel angle:

$$r = \frac{\Psi d}{\chi \cos \left(\frac{1}{2} \left(\chi - \Psi \right) \right)} \quad \text{with: } \chi = \pi - \Phi .$$

The gray area in Figure 2 is a 2D representation of the billet with the length h and the width d, which is equal to the diameter of the entrance channel. The geometry parameters for the 2D model presented in Table 1 are chosen in analogy to the parameters used for the experimental investigations shown later.

Table 1. Geometry parameters for the simulation model (cf. Figure 2).

Φ	Ψ	R	d	d_{ex}	h	l_{en}	l_{ex}	l_1	l_2
90°	1°	$5r$	50 mm	52 mm	300 mm	300 mm	75 mm	5 mm	41 mm

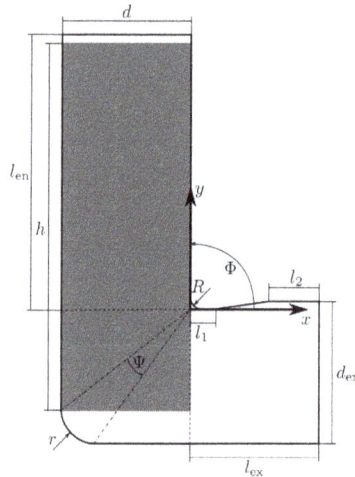

Figure 2. Sketch of the simulation model with the relevant geometrical parameters.

To simulate the ECAP process, a version of the elastic visco-plastic material model of Shutov and Kreißig [30] is used (cf. Table 2). It is based on the multiplicative decomposition of the deformation gradient $\underline{F} = \underline{F}_i \cdot \underline{F}_e$ and of the inelastic part of the deformation gradient $\underline{F}_i = \underline{F}_{ii} \cdot \underline{F}_{ie}$. The rheological representation of the model [31] is shown schematically in Figure 3.

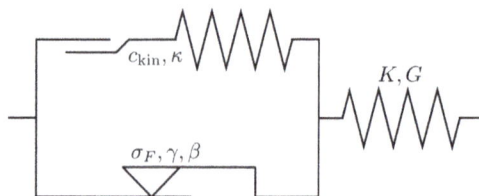

Figure 3. Rheological representation of the material model with the material parameters.

The elastic behavior is modeled with a Neo-Hookean spring and the plastic flow is represented by a parallel connection of a St. Venant element with an endochrone element that describes the kinematic hardening of an Armstrong-Frederick type [32]. Furthermore, the St. Venant element incorporates the yield stress and the isotropic hardening of Voce type [33]. Normally, the visco-plastic flow is prescribed with Perzyna's rule [34]. However, in the present study, rate-independent plastic flow is assumed.

Table 2. Constitutive equations of the applied elastic-plastic material model.

stress:	$\widetilde{\underline{T}} \cdot \underline{C} = K \ln\left(I_3\left(\underline{C}\right)\right) \underline{I} + G \left(\underline{C}_i^{-1} \cdot \underline{\tilde{C}}\right)^D$	(1)
flow rule:	$\underline{C}_i^{-1} \cdot \overset{\triangle}{\underline{C}}_i = \dfrac{\lambda}{\mathfrak{F}} \left(\widetilde{\underline{T}} \cdot \underline{C} - \widetilde{\underline{T}}_i \cdot \underline{C}_i\right)^D$	(2)
	with: $I_3\left(^0\underline{C}_i\right) = 1$ and $\underline{C}_i|_{t=t_0} = {}^0\underline{C}_i$	
kinematic hardening:	$\widetilde{\underline{T}}_i \cdot \underline{C}_i = \dfrac{c_{kin}}{2} \left(\underline{C}_{ii}^{-1} \cdot \underline{C}_i\right)^D$	(3)
	$\underline{C}_{ii}^{-1} \cdot \overset{\triangle}{\underline{C}}_{ii} = \dfrac{1}{2}\lambda\kappa c_{kin} \left(\underline{C}_{ii}^{-1} \cdot \underline{C}_i\right)^D$	(4)
	with: $I_3\left(^0\underline{C}_{ii}\right) = 1$ and $\underline{C}_{ii}|_{t=t_0} = {}^0\underline{C}_{ii}$	
flow function:	$\Phi = \mathfrak{F} - \sqrt{\dfrac{2}{3}}\left(\sigma_{F0} + R\right)$	(5)
equivalent stress:	$\mathfrak{F} = \mathfrak{N}\left(\left(\widetilde{\underline{T}} \cdot \underline{C} - \widetilde{\underline{T}}_i \cdot \underline{C}_i\right)^D\right)$	(6)
	with: $\mathfrak{N}\left(\underline{X}\right) := \sqrt{X_{ab} X_{ba}}$	(7)
isotropic hardening:	$\dot{R} = \dot{s}\left(\gamma - \beta R\right)$ with: $\dot{s} = \sqrt{\dfrac{2}{3}}\lambda$	(8)
effective plastic strain:	$s(t) = \displaystyle\int_{\tau=0}^t \dot{s}(\tau)\,d\tau$	(9)
KKT conditions:	$\lambda \geq 0, \quad \Phi \leq 0, \quad \lambda\Phi = 0$	(10)

Note that the constitutive model is a purely phenomenological elastic-plastic model with isotropic and kinematic hardening. It is not based on specific assumptions regarding microstructural deformation mechanisms and therefore allows for a quite general analysis of shear localization during ECAP. As indicated in the constitutive Equations (1)–(10), seven parameters are needed to describe the material behavior. K and G are the bulk and shear modulus, σ_{F0} is the yield stress and the other constants represent the parameters of the isotropic (β, γ) and kinematic (κ, c_{kin}) hardening. As shown in [35], β and κc_{kin} define the initial hardening rate (IHR) while $\frac{\gamma}{\beta}$ and κ^{-1} determine the strain hardening capacity (SHC). (The meaning of the kinematic hardening parameters is derived from linearized equations. Strictly speaking, this interpretation is thus only valid for small strains.)

In the material model summarized in Table 3, the Cauchy-Green tensor $\underline{C} = \underline{F}^T \cdot \underline{F}$, two internal state variables $\underline{C}_i$, $\underline{C}_{ii}$ and the second Piola-Kirchoff stress $\widetilde{\underline{T}}$ further appear, which is related to the Cauchy stress $\underline{\sigma}$ by

$$\widetilde{\underline{T}} = I_3\left(\underline{F}\right) \underline{F}^{-1} \cdot \underline{\sigma} \cdot \underline{F}^{-T}.$$

Here, the third invariant I_3 of a second order tensor $\underline{X} = X_{ab}\,\underline{e}_a \otimes \underline{e}_b$ is defined by $I_3(\underline{X}) = \det[X_{ab}]$. Furthermore, $\overset{\triangle}{\underline{X}}$ denotes the material time derivative and $\underline{X}^D$ represents the deviatoric part of $\underline{X}$. Finally, $\underline{\bar{C}} = I_3\left(\underline{C}\right)^{-\frac{1}{3}} \underline{C}$ denotes the unimodular part of $\underline{C}$. The identity tensor is prescribed as the initial condition for the internal state variables $\left({}^0\underline{C}_i = \underline{I}, {}^0\underline{C}_{ii} = \underline{I}\right)$ throughout this paper. The inelastic multiplier λ is calculated iteratively from the Karush–Kuhn–Tucker (KKT) conditions (10) applying a Newton iteration.

The material model was implemented in the VUMAT user subroutine of Abaqus/Explicit. Further details on the material model and applications to the simulation of metal forming can be found elsewhere [30,36–38]. For an adequate interpretation of the results, some simplifications in the simulation model must be kept in mind:

1. A two-dimensional plane-strain simulation is used. Therefore, boundary effects perpendicular to the simulated plane have to be neglected. The main reason for this simplification is the computational effort, which increases excessively when changing to a three-dimensional model. However, the comparison of simulation and experimental results (Section 3) demonstrates that the error of this simplification is small.
2. As the investigation of viscous effects goes beyond the scope of this study, a rate-independent model was chosen. The aluminum material used in the experimental studies exhibits only a minor rate dependence and the experiments have been performed at room temperature. Due to the very low pressing speed of 0.3 mm/s and consequently low strain rates inside the shear zone, the error of this limitation is kept small. Furthermore, adiabatic heating phenomena are essentially negligible and do not have any practical significance [39].
3. A frictionless model was used to simulate ECAP. In the experimental studies, the ECAP die is always lubricated such that the friction coefficient is low [40]. In addition, both in the experiments and in the simulations, the exit channel is moved with the billet in order to minimize the friction [41]. Because of this, self-heating due to friction may be neglected.

3. Parameter Variation

First, convergence studies for element type, element size, mass scaling and velocity scaling were performed to reach a behavior independent of these influences and to avoid artificial effects (e.g., due to inertia forces). The FE model was meshed with rhomboid-shaped CPE4R-elements. The initial inclination was chosen such that the elements' interior angles remain as close as possible to 90° during the whole deformation. It turned out that a mesh with a maximum element size of $e = 0.5$ mm ($\approx$92,600 elements) has to be used by applying a maximum possible mass scale of $f_m = 400$ and plunger velocity of $v = 0.75$ m/s. After these convergence studies, a parameter variation for almost all geometrical and material parameters was conducted. Here, only the most important results are presented; for a more detailed depiction including the convergence studies, see [35].

3.1. Reference Simulation

As a reference, a single ECAP pass is simulated with elastic ideal-plastic behavior. The material parameters corresponding to aluminum are shown in Table 3.

Table 3. Material parameters for aluminum with ideal-plastic behavior, used in the reference simulation.

K/MPa	G/MPa	σ_{F0}/MPa	β	γ/MPa	c_{kin}/MPa	κ/MPa^{-1}
73,500.0	28,200.0	270.0	0.0	0.0	0.0	0.0

In the contour plot in Figure 4, some inhomogeneities in the strain field are visible. However, because of the low difference between maximum and minimum value and the missing regularity, they do not represent fully developed shear or matrix bands (see Section 1). Instead, if the amplitude of the effective plastic strain is lower than 0.15, this state will be called "minor fluctuation" in the following. Furthermore, the outer edge of the ECAP die is completely filled with material (lower left corner in Figure 4), which is in line with experimental investigations [23]. Finally, there is a small curvature of the billet because of internal stresses developed during ECAP [42]. This is also known from ECAP experiments *without* applying any backpressure. In summary, the simulation result corresponds well with the expectations gained from experiments.

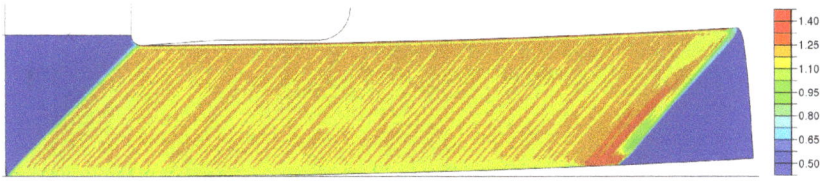

Figure 4. Contour plot of the effective plastic strain *s* (cf. Equation (9)) for the reference simulation: some inhomogeneities are visible. Compared to experimental observations of shear and matrix bands, these variations are negligible; they are therefore considered as minor fluctuations.

3.2. Effects of Isotropic Hardening

Strain localization is observed in experiments if the SHC of the material subjected to ECAP is low. Motivated by these experimental observations, two different levels of the hardening capacity: $\frac{\gamma}{\beta} = 15$ MPa and $\frac{\gamma}{\beta} = 200$ MPa and three different levels of the initial hardening rate (IHR): $\beta = 0.1$, $\beta = 1.0$ and $\beta = 10.0$ are chosen for a systematic comparison. For $\beta = 0.1$, the material has a low and for $\beta = 1.0$ it has a high hardening rate during the whole deformation. For $\beta = 10.0$, the material has a pronounced IHR and a very small hardening rate at the end of the deformation.

The results of the simulations with $\frac{\gamma}{\beta} = 15$ MPa and with $\frac{\gamma}{\beta} = 200$ MPa are shown in Figures 5 and 6, respectively. Minor strain fluctuations occur for all simulations, except the one with a high SHC and a high hardening rate during the whole deformation, Figure 6b. In this case, even minor fluctuations are suppressed yielding a quasi-homogeneous distribution. The situation is different for the low SHC ($\frac{\gamma}{\beta} = 15$ MPa): even the high hardening rate during the whole deformation is still too low to suppress fluctuations. Consequently, the SHC controls the heterogeneity of plastic deformation in case of isotropic hardening. However, the difference between the ideal-plastic behavior and this hardening behavior is very small and therefore the results are similar.

For $\frac{\gamma}{\beta} = 200$ MPa, the SHC is much higher. Still, for low IHR, the material behavior is almost ideal-plastic as well and, due to this, fluctuations occur. For high IHR, the material is hardening during the whole deformation and any heterogeneities are suppressed. For pronounced IHR, the SHC is already exhausted at the initial stage of the deformation process. Thus, at the end of the deformation, the material behaves almost ideal-plastically with a higher yield stress. For this reason, heterogeneities also occur in this case.

There is another feature that becomes visible looking at the bottom zone of the billets shown in Figures 5 and 6: the more the material flow omits the outer edge of the channel, the less the bottom zone deforms and the higher the vertical gradient of the effective plastic strain and the billet's curvature. This detail is in perfect agreement with experimental observations [41,43,44], which further indicates that the FE simulations presented here accurately capture various subtle experimental details.

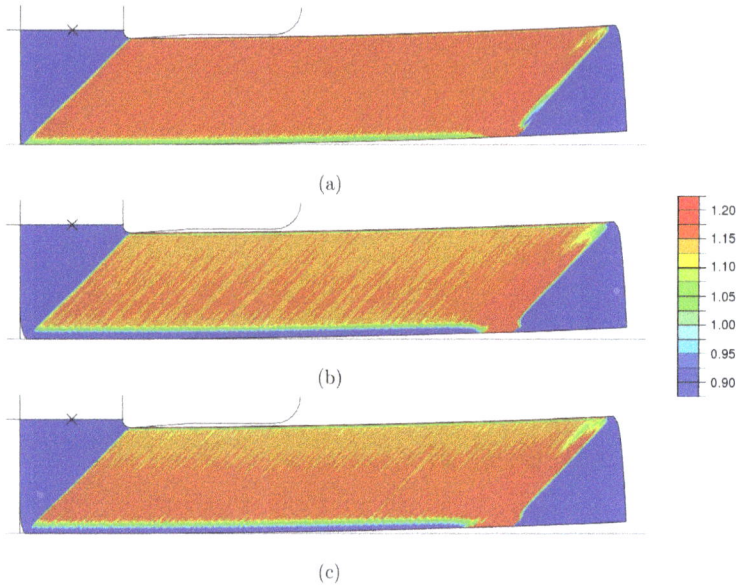

Figure 5. Effective plastic strain s (cf. Equation ($\cup$)) for simulations with isotropic hardening with $\frac{\gamma}{\beta} = 15$ MPa in combination with (**a**) $\beta = 0.1$, (**b**) $\beta = 1.0$ and (**c**) $\beta = 10.0$. $\sigma_F = 270$ MPa. In all cases, only minor fluctuations occur.

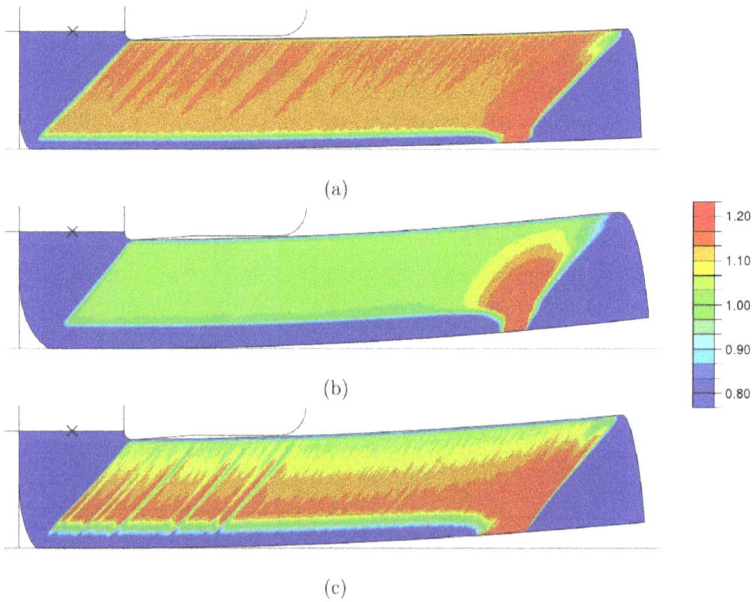

Figure 6. Effective plastic strain s (cf. Equaiton ($\cup$)) for simulations with isotropic hardening with $\frac{\gamma}{\beta} = 200$ MPa in combination with (**a**) $\beta = 0.1$, (**b**) $\beta = 1.0$ and (**c**) $\beta = 10.0$. $\sigma_F = 270$ MPa. In (**a**,**c**), minor fluctuations occur; (**b**) shows a quasi-homogeneous distribution of the effective plastic strain.

3.3. Effects of Kinematic Hardening

The difference between kinematic and isotropic hardening is the plastic anisotropy induced by kinematic hardening. While modeling of isotropic hardening usually involves only the *scalar* equivalent stress, kinematic hardening is based on the stress and back stress *tensor* (cf. Equation (3)). Hence, kinematic hardening not only depends on the principal stresses, but also on the principal *directions*.

A high SHC of $\kappa^{-1} = 200$ MPa was chosen and the IHR was varied in a similar way as in the investigations for isotropic hardening. The results shown in Figure 7 are similar to those of isotropic hardening. At low IHR and at pronounced IHR strain, inhomogeneities occur, whereas, at high IHR, the inhomogeneities are suppressed. However, there is an important difference: at pronounced IHR, the deformation is not only inhomogeneous but, as an entirely new feature, a very regular pattern emerges. Additionally, the difference between the minimum and maximum strain values in the distinctly different regions is high ($\Delta s \approx 1.0$).

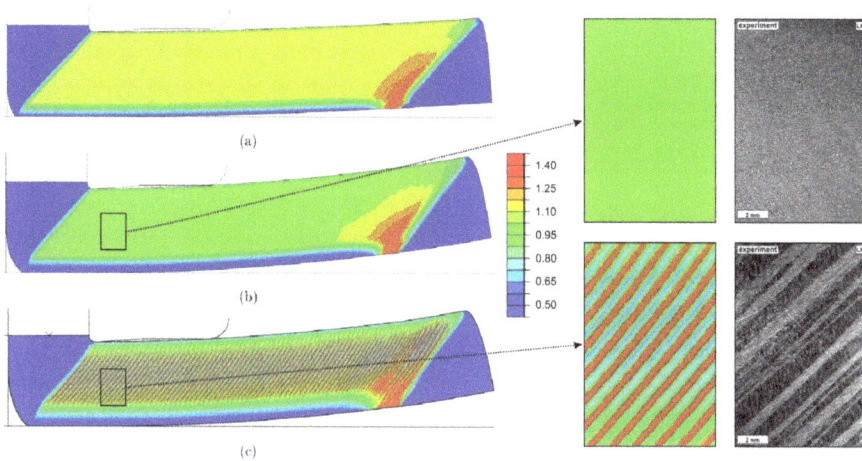

Figure 7. (**left**) effective plastic strain s (cf. Equation (9)) for simulations with kinematic hardening with $\kappa^{-1} = 200$ MPa in combination with (**a**) $\kappa c_{kin} = 0.075$, (**b**) $\kappa c_{kin} = 0.25$ and (**c**) $\kappa c_{kin} = 3.75$; (**right**) comparison of the simulations with optical micrographs after one pass of ECAP for a fully recrystallized (**top**) and a severely cold worked conventional 6000 series aluminum alloy (**bottom**).

This type of localized deformation is very similar to the shear and matrix band structure observed in the experiment (cf. Section 3.1). The regularity of these bands can also be documented by plotting the progression of the effective plastic strain along the length of the billet, Figure 8. The undeformed regions of the billet (i.e., the material near the ends of the billet that has not passed through the shear zone during ECAP) and the region of unsteady flow at the beginning of the process are omitted in this analysis; the total length of the billet is 300 mm. Furthermore, an animation of this simulation is provided in the supplementary material of this paper.

To ensure a converged solution, the simulation model of the billet was divided into five parts (cf. Figure 9). The two big outer parts are meshed with an element size of $e = 0.5$ mm, the two parts next to them with an element size of $e = 0.25$ mm and the part in the midst with an element size of $e = 0.125$ mm. Despite the mesh size variation with the length of the billet, the shear and matrix bands dimensions remain constant and some shear bands cross the border between two mesh parts. This shows that the dimensions and the direction of the shear bands do not depend on the mesh. The difference in the effective plastic strain increases with a further refinement of the mesh.

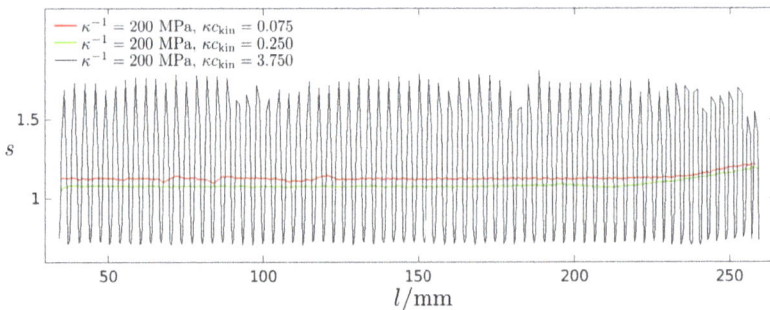

Figure 8. The effective plastic strain s (cf. Equation (9)) as a function of the position in the billet (cf. Figure 7). Undeformed areas and the unsteady flow at the beginning of the process are omitted.

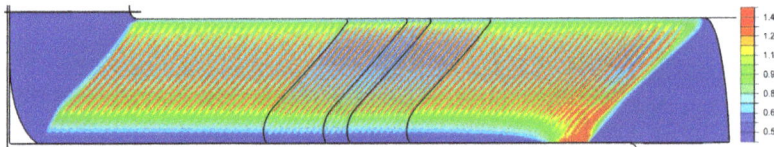

Figure 9. Contour plot of the effective plastic strain s (cf. Equation (9)) for the simulation with locally refined mesh. The billet is divided into five parts. The outer two parts are meshed with an element size of $e = 0.5$ mm, the two parts next to them with an element size of $e = 0.25$ mm and the part in the midst with an element size of $e = 0.125$ mm. Obviously, the solution is converged with respect to the dimensions of the shear and matrix bands.

In order to further analyze and confirm the observation that shear bands are formed when there is a pronounced IHR, additional simulations were carried out, varying κ and c_{kin}. In Figure 10, three examples are shown: in simulation (a), c_{kin} was doubled resulting in an increase of the IHR (cf. Section 2). With this increase of the IHR, many more shear bands are formed. In simulation (b), κ^{-1} is halved, which decreases the SHC and reduces the IHR. This also increases the number of shear bands, but the difference in effective strain between shear and matrix bands is considerably decreased ($\Delta s \approx 0.2$). In simulation (c), κ^{-1} and c_{kin} are halved such that the hardening capacity is lower, but the IHR remains unchanged. In this case, the number of shear bands increases as well compared to Figure 7c. Based on these (and many other performed) simulations, it can be concluded that there is a strong effect of the kinematic hardening parameters on the emergence, number, and shape of the bands as well as on the amount of plastic strains within them. However, this effect is strongly nonlinear and the occurrence of distinct shear bands cannot be simply related to a single material parameter in the framework of the present material model.

Finally, one key result of the present study is that—while fluctuations in terms of the distribution of effective strains along the billet can in principle be simulated both by assuming isotropic or kinematic hardening, and using a wide variety of material parameters—only kinematic hardening appears to be suitable for an accurate simulation of distinct shear bands that differ from the adjacent matrix bands by relatively large amounts of plastic strain.

The results may be summarized in a generalized diagram where the parameter regions are indicated with corresponding uniaxial flow curves as function of s (Figure 11). (Note that the diagram is only valid for reasonable i.e., sufficiently high values of the SHC). Heterogeneities occur in two different regions, defined approximately by the solid lines: in the region where the hardening rate is too low during the whole deformation such that the material behavior is similar to ideal-plastic, as well as in the region where the maximum hardening capacity is reached shortly after the beginning of the deformation. For isotropic hardening (cf. Figure 11a), all heterogeneities remain minor fluctuations whereas for kinematic hardening (cf. Figure 11b) shear and matrix bands are formed. The region defined approximately by the dotted curves corresponds to the quasi-homogeneous solution.

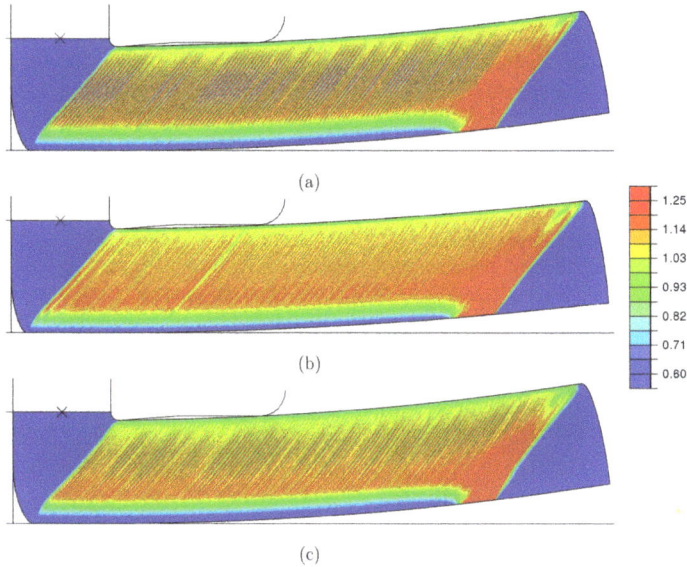

Figure 10. Effective plastic strain s (cf. Equation (9)) for simulations with kinematic hardening with the parameters (a) $\kappa^{-1} = 200$ MPa, $\kappa c_{kin} = 7.5$, (b) $\kappa^{-1} = 100$ MPa, $\kappa c_{kin} = 7.5$ and (c) $\kappa^{-1} = 100$ MPa, $\kappa c_{kin} = 3.75$. In all cases, shear and matrix bands occur.

Remark 1. *Only the reference simulation (and not a solution with such strong heterogeneities as in Figures 7c, 9 and 10) was considered in the convergence studies. However, here a converged solution with respect to the number of the shear and matrix bands is presented. While the number of shear and matrix bands stays constant when the mesh size is further reduced, the values of the effective strain in the bands have not fully converged yet.*

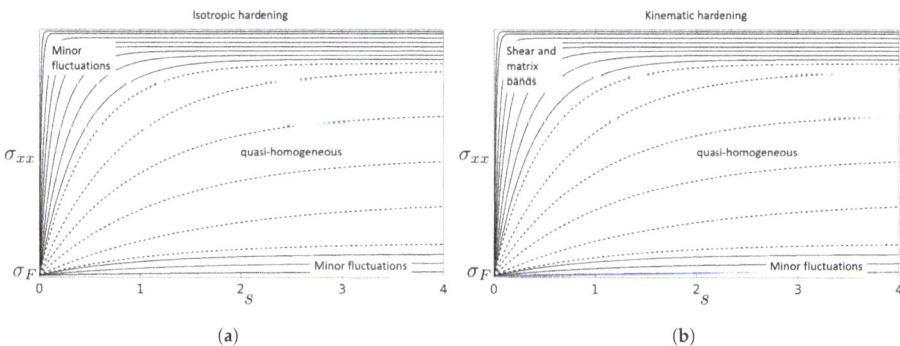

Figure 11. Generalized diagram with uniaxial flow curves as functions of the effective plastic strain depicting the regions where localized deformation occurs for (a) isotropic and (b) kinematic hardening. Material parameters corresponding to flow curves in the marked regions lead to the indicated solution. Note that, despite the apparent similarity of the *uniaxial* flow curves, isotropic and kinematic hardening are fundamentally different in the *multiaxial* case.

3.4. Experimental Validation

In order to validate the simulation results, complementary experimental studies were carried out using technical parameters where shear localization has been reported before. It is well known that ECAP typically introduces a homogeneous shear deformation for many metallic materials [9]. One main requirement for a homogeneous deformation is a sufficient SHC of the processed material. In the case of processing conventional fully recrystallized 6000 series aluminum alloys, a homogeneous introduction of strain is generally observed [43–45]. Figure 12 shows the microstructure of an ECAP-processed billet (AA6060) that has been processed conventionally (hot extrusion including full recrystallization) prior to ECAP. Figure 12a shows an optical micrograph of the severely sheared microstructure without any macroscopic heterogeneities. A detailed view into the microstructure by scanning electron microscopy (in electron back-scatter diffraction mode, EBSD) reveals minor differences regarding the locally introduced strains that result from the different crystallographic orientations of Figure 12b,c. These microstructural results are completely in line with earlier reports [9,46,47] showing macroscopically homogeneous deformation for similar alloys.

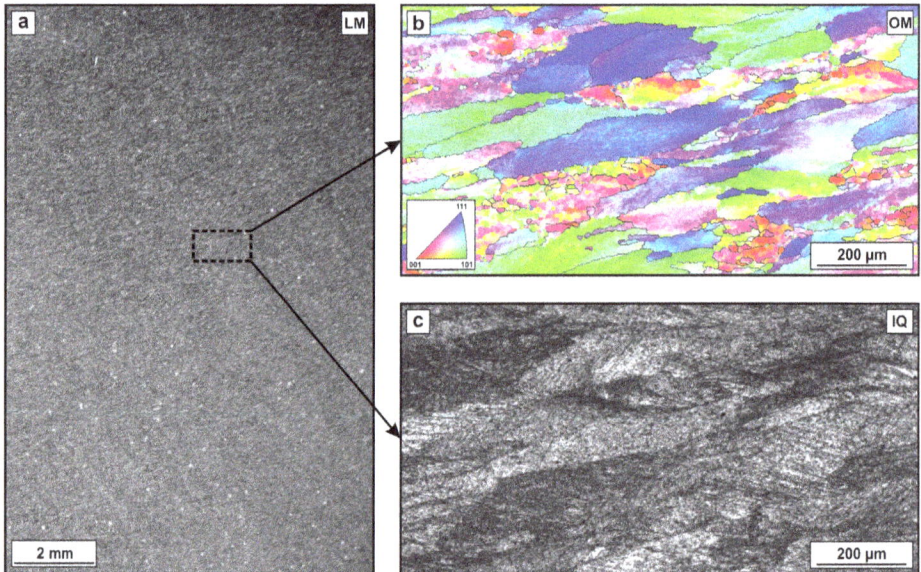

Figure 12. Microstructure after homogeneous shear deformation by one pass of ECAP of a fully recrystallized conventional 6000 series aluminum alloy. Optical micrograph (**a**) showing a microstructural overview and color-coded orientation map (OM) (**b**) and image quality map (IQ) (**c**) from electron back-scatter diffraction (EBSD) measurement.

These results were obtained for a fully recrystallized material that exhibits a sufficiently high hardening rate throughout the whole deformation process. The analogous simulation parameters in the conducted simulations for isotropic hardening are $\frac{\gamma}{\beta} = 200$ MPa, $\beta = 1.0$ and for kinematic hardening $\kappa^{-1} = 200$ MPa, $\kappa c_{kin} = 0.25$. Both simulations result in a quasi-homogeneous solution, as shown for isotropic hardening in Figure 6b and for kinematic hardening in Figure 7b. Thus, a very good accordance between simulation and experiments is achieved for this kind of material condition.

In contrast to the homogeneous microstructure observed after ECAP of a fully recrystallized material, a strongly heterogeneous structure was found after processing a severely cold-worked condition (cold extrusion [48]) of the same alloy. As a consequence of cold extrusion the material exhibits a very limited SHC compared to its hot-extruded counterpart. Figure 13a shows the microstructure after ECAP. It is characterized by an alternating arrangement of two fundamentally different types of macroscopic bands: Shear bands, where the material got severely sheared and matrix bands that exhibit much lower strains. Figure 13c shows an image quality map of an EBSD measurement. Darker areas are associated with a low band contrast as a result of severe distortion of the lattice due to inherent strain. The band structure is oriented under an angle of exactly 45° with respect to the pressing direction, which corresponds to an alignment parallel to the theoretical shear zone [3,49] of the ECAP die. The mean width of both band types is approximately 400 μm resulting in an almost equal area fraction (≈50%). It should be noted that the material might show a pronounced kinematic strain hardening caused by the previous cold working by extrusion at ambient temperature.

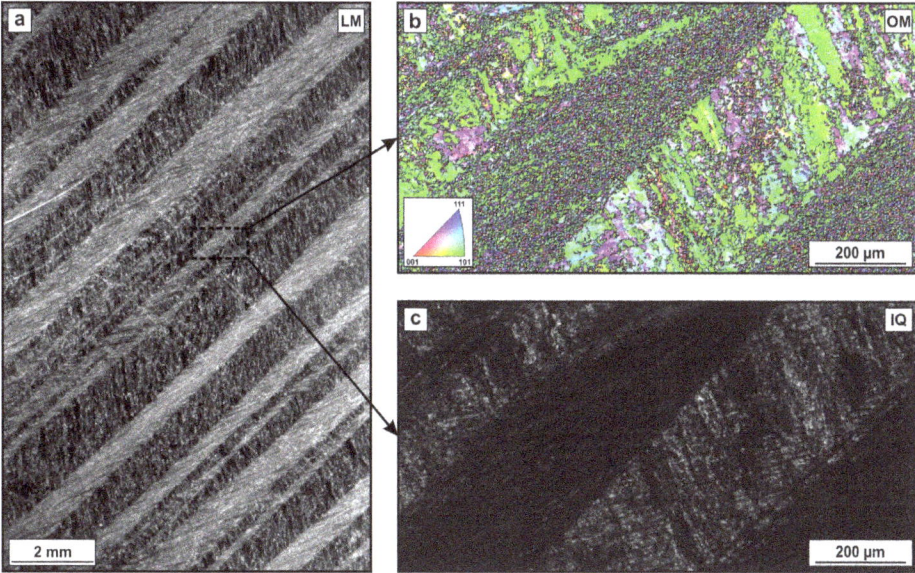

Figure 13. Microstructure after heterogeneous deformation by one pass of ECAP of a severely cold worked conventional 6000 series aluminum alloy. Optical micrograph (**a**) showing a microstructural overview and color-coded orientation map OM (**b**) and image quality map IQ (**c**) from EBSD measurement.

The material behavior in this case corresponds to the simulations for isotropic hardening with $\frac{\gamma}{\beta} = 200$ MPa, $\beta = 10.0$ or for kinematic hardening with $\kappa^{-1} = 200$ MPa, $\kappa c_{kin} = 3.75$. These simulations show a very small SHC during the entire ECAP pass (cf. Remark 2). As shown in Figure 7, there exists a very good agreement between the result of the simulation with kinematic hardening and the experimental strain distributions of shear and matrix band. (An explanation of why a simulation including only isotropic hardening cannot produce shear and matrix bands is given in Section 4.) Consequently, a scale comparison was conducted in Figure 14, which also shows excellent agreement. The length scales of the shear and matrix bands in the experiments and in the simulations are nearly equal.

Remark 2. *To be more precise, these simulations show at the beginning of the process a high SHC, but also a pronounced hardening rate. With this particular material behavior, the hardening of the material during the extrusion prior to ECAP can be captured. However, after a minimal plastic deformation in the simulation, the material already reaches the state of a very small remaining SHC.*

Despite the inherent simplifying assumptions (cf. Section 2) and even though the experiments are affected by various technological influencing factors like friction, there is a close agreement between the simulations and the experimental observations. This encourages and motivates further analysis of the microstructural and mechanical mechanisms that lead to the formation of shear bands during ECAP. A first attempt to more generally understand which conditions lead to distinct shear localization is given below.

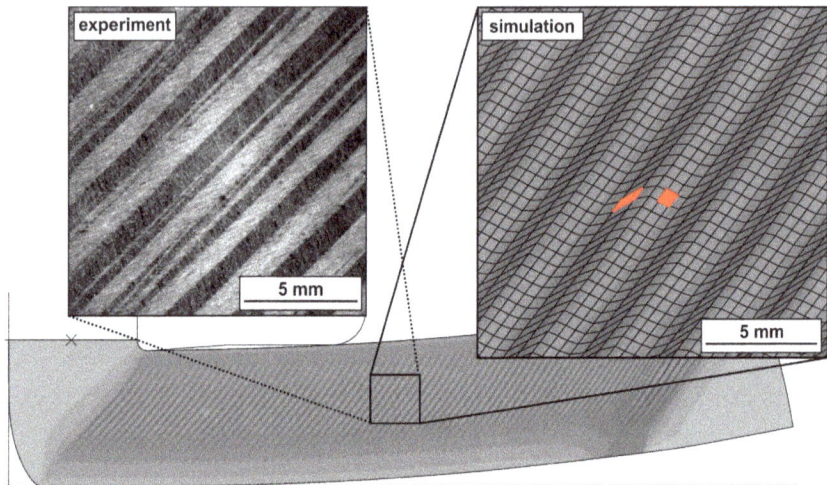

Figure 14. Comparison of shear and matrix bands in simulation (**right**) and experiment (**left**). In the optical micrograph, the shear bands are shown in bright and the matrix bands in dark gray. In the simulation, the opposite is the case: The shear bands are depicted in dark and the matrix bands in bright gray. Additionally, the positions of the chosen elements in the billet far away from all boundaries are shown in red.

4. Remarks on the Mechanism of Heterogeneous Plastic Flow

To get a more detailed description of the localization mechanism, two representative finite elements are chosen: one element of a shear and one element of a matrix band, as depicted in Figure 14 (simulation). These elements are selected far away from all boundaries of the ECAP tool, such that no boundary effects affect the results. For these elements, the stress and strain state is evaluated.

At first, the evolution of stresses and strains during one ECAP pass in the shear and matrix bands is investigated. Shear and matrix band evolve one after another and there is a constant time interval between the formation of these bands. For reasons of clarity and comprehensibility, the time interval is subtracted in every diagram comparing shear and matrix bands (Figures 15 and 16). As shown in Figure 15, the evolution is equal for both types of bands for a long time. However, at a certain point after the plastic flow has already set in, the solution diverges abruptly. At this point of the deformation, the effective plastic strain rate $\dot{s}$ increases rapidly in the shear bands, whereas, in the matrix bands, $\dot{s}$ decreases. It follows that, at this point, the plastic flow increases rapidly in the shear band as the rate of plastic deformation is directly proportional to the effective plastic strain rate: $(\dot{\lambda} = \sqrt{3/2}\dot{s})$. Interestingly, the duration of plastic flow is equal for both bands, as shown in the yellow domain in Figure 15. This is in contrast to results given in the literature [11], where it is assumed that plastic flow occurs *only* in the shear bands. The simulation results show that also in the matrix bands plastic flow occurs while passing through the shear zone, even though to a very small extent. The simulations also indicate that the deformation of both types of bands occurs in the same shear zone region within the ECAP die.

To reach a better comprehension of the evolution of the shear and matrix bands, the evolution of the corresponding stress states is also considered. Figure 16a compares the von Mises stress of the shear and the matrix band. Interestingly, the von Mises equivalent stress remains almost equal until far after the divergence of the effective plastic strain. For a justification of this fact, the definition of the von Mises stress

$$\sigma_{vM} = \sqrt{\frac{1}{2}\left[(\sigma_{xx} - \sigma_{yy})^2 + (\sigma_{xx} - \sigma_{zz})^2 + (\sigma_{zz} - \sigma_{yy})^2\right] + 3\left(\sigma_{xy}^2 + \sigma_{xz}^2 + \sigma_{yz}^2\right)} \tag{11}$$

has to be considered. It is clear that equal normal stresses are not required for an equal von Mises stress; only the difference between the coefficients of the stress tensor has to be equal. This observation gives an explanation why shear bands only appear by involving kinematic hardening and not by isotropic hardening: isotropic hardening is based on the equivalent stress (cf. Equation (8)). If there is no difference between shear and matrix band in terms of the equivalent stress, there will be no differences in material behavior. Therefore, there is no difference concerning effective strain.

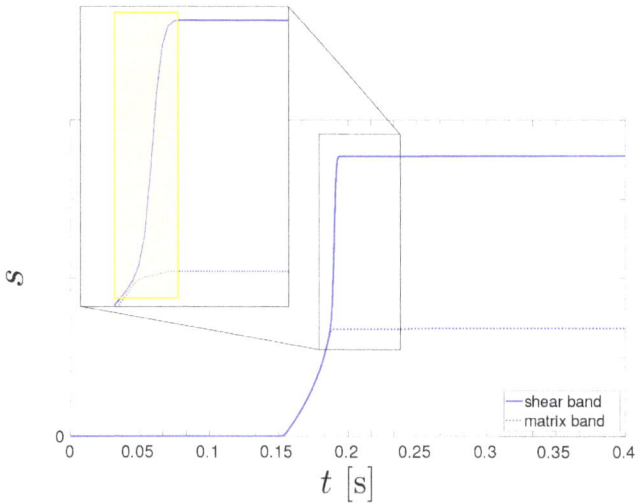

Figure 15. The evolution of the effective plastic strain *s* (cf. Equation (9)) for the representative elements corresponding to shear and matrix bands, respectively (see highlighted elements in Figure 14).

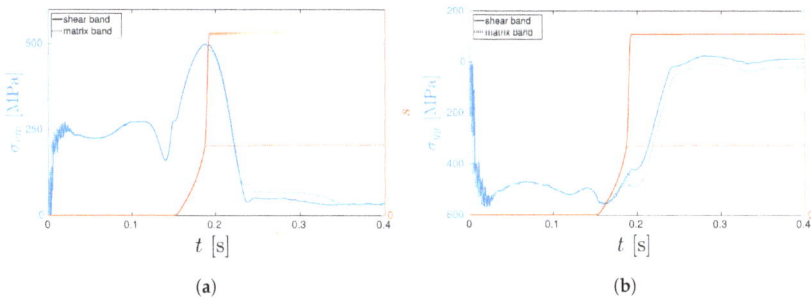

(a)   (b)

Figure 16. The evolution of the effective plastic strain *s* (orange) and of (**a**) the von Mises equivalent stress (blue) and of (**b**) the principal stress in ECAP pressure direction (blue). The solid lines are used for the shear band and the dotted lines for the matrix band.

As a consequence of almost equal von Mises stresses, all components of the stress tensor have to be analyzed. The focus here is placed on the normal component with respect to the pressing direction y. (cf. Figure 2—Note that the y-direction is defined globally. Hence, it is only the pressing direction for the input channel.) Again, the evolution of σ_{yy} within shear and matrix bands remains very similar for a long time. As indicated in Figure 16b, a difference in σ_{yy} occurs just in time with the difference in the effective plastic strain within shear and matrix bands. The fact that it is difficult to identify a precise source/origin of the divergence of the solution into shear and matrix bands is typical for pattern-forming systems. In such systems, assigning cause and effect is not always directly possible [50,51]. Moreover, one peculiarity becomes obvious: the difference in stress and strain state emerges abruptly and there is no indicator for a slowly increasing gap between the bands. This indicates an analogy to a supercritical pitchfork bifurcation in the solution (cf. Figure 17). The control parameter in such a theoretical scenario is likely to be a combination of c_{kin} and κ, as shown in Section 3. There are domains in the parameter space where only one stable solution occurs, the homogeneous distribution of strain. However, in other domains, two stable solutions occur: the shear and the matrix band. The choice of the stable solution is affected by the previously deformed material segment. A material segment that experiences an overlarge plastic flow (the shear band) is followed by a material segment that experiences an undersized plastic flow (the matrix band). It is worth noting that the application of bifurcation analysis (which is common in studying pattern-forming systems) to ECAP represents a novel approach to understand the formation of shear and matrix bands. While the fundamental idea is presented here for the first time, it is highlighted that further work is required to fully describe and understand localized flow during ECAP in the conceptual framework of pattern formation and self-organizing systems.

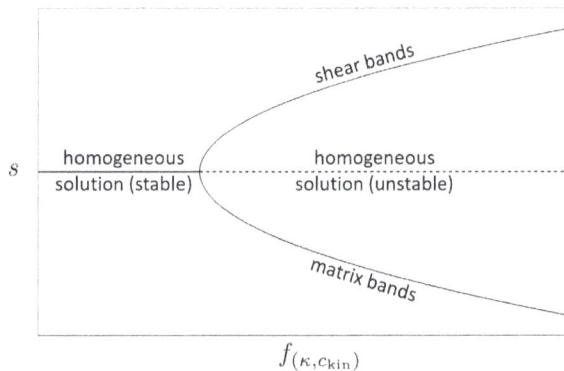

Figure 17. Assumed supercritical pitchfork bifurcation of the effective plastic strain s (cf. Equation (9)) in the parameter space. The control parameter is a function of both parameters of kinematic hardening.

5. Conclusions

Adopting phenomenological plasticity theory with isotropic and kinematic hardening and using two-dimensional explicit finite element simulations, and strain localization during ECAP can be reproduced. To this end, a systematic convergence study helps to ensure a stable and reliable FE solution, which is suitable for gaining new insights in the localization process. The FE simulations show that isotropic hardening can only cause minor fluctuations in the plastic strain fields. Kinematic hardening with high initial hardening rate and sufficient strain hardening capacity can lead to pronounced localized deformation in form of shear and matrix bands. This enables the numerical analysis of the evolution of this band structure during ECAP with a phenomenological material model. Neither micro-mechanical material features nor strain softening of the material have to be considered. It is confirmed that kinematic harding plays thus an important role for simulating strain localization during ECAP. Additionally, we find a surprisingly accurate match between simulation and our experimental results. A detailed analysis of shear and matrix bands revealed that not only the shear bands, but also the matrix bands deform plastically during ECAP. Although the plastic deformation is very small in the whole shear zone, the effective plastic strain rate is always larger than zero.

In future work, the influences of the different hardening mechanisms have to be studied in greater detail. In particular, the effect of the parameters of kinematic hardening but also the influence of the formative hardening on the evolution of shear bands is not fully understood yet. An extensive experimental characterization of the present hardening behavior of the investigated material is the focus of a future work. This will help to gain a detailed understanding of the relationship between material's hardening behavior and the occurrence of strain localization. Because of the complex relationship of strain hardening and local (plastic) deformation behavior including strain localization in certain cases, an in-depth understanding of the acting microstructural and micro-mechanical mechanisms is needed to describe and predict the material's behavior during severe plastic deformation. Applying the theoretical concepts used to describe pattern-formation in self-organizing systems may provide a novel pathway to distinguish stable vs. unstable deformation modes.

Supplementary Materials: The following are available online at www.mdpi.com/2075-4701/8/1/55/s1, Video S1: Effective strain for kinematic hardening with the hardening parameters $\kappa^{-1} = 200$ MPa, $\kappa c_{kin} = 3.75$. The shear and matrix band formation in the shear zone is shown.

Acknowledgments: The authors gratefully acknowledge the German Research Foundation (Deutsche Forschungsgemeinschaft, DFG) for supporting this work carried out within the framework of the Collaborative Research Center SFB 692 (projects A1, C2, C5).

Author Contributions: Tobias Daniel Horn performed the FE simulations, based on a pilot study from Christian Silbermann. Philipp Frint conceived and designed the experiments. All authors discussed the results.

Conflicts of Interest: The authors declare no conflict of interest.

Abbreviations

The following abbreviations are used in this manuscript:

ECAP	equal-channel angular pressing
SPD	severe plastic deformation
IHR	initial hardening rate
SHC	strain hardening capacity
EBSD	electron back-scatter diffraction
KKT	Karush–Kuhn–Tucker
FE	finite element

References

1. Segal, V. The Method of Material Preparation for Subsequent Working. Patent of the USSR Nr. 575892, 1977.
2. Segal, V.; Reznikov, V.; Dobryshevshiy, A.; Kopylov, V. Plastic working of metals by simple shear. *Russ. Metall. (Metally)* **1981**, *1*, 99–105.
3. Iwahashi, Y.; Wang, J.; Horita, Z.; Nemoto, M.; Langdon, T.G. Principle of equal-channel angular pressing for the processing of ultra-fine grained materials. *Scr. Mater.* **1996**, *35*, 143–146.
4. Hall, E. The deformation and ageing of mild steel: III discussion of results. *Proc. Phys. Soc. Sect. B* **1951**, *64*, 747.
5. Petch, N. The cleavage strength of polycrystals. *J. Iron Steel Inst.* **1953**, *174*, 25–28.
6. Valiev, R.; Alexandrov, I.; Zhu, Y.; Lowe, T. Paradox of strength and ductility in metals processed bysevere plastic deformation. *J. Mater. Res.* **2002**, *17*, 5–8.
7. Wang, Y.; Chen, M.; Zhou, F.; Ma, E. High tensile ductility in a nanostructured metal. *Nature* **2002**, *419*, 912–915.

8.   Ma, A.; Jiang, J.; Saito, N.; Shigematsu, I.; Yuan, Y.; Yang, D.; Nishida, Y. Improving both strength and ductility of a Mg alloy through a large number of ECAP passes. *Mater. Sci. Eng. A* **2009**, *513*, 122–127.

9.   Valiev, R.Z.; Langdon, T.G. Principles of equal-channel angular pressing as a processing tool for grain refinement. *Prog. Mater. Sci.* **2006**, *51*, 881–981.

10. Kim, W.; Chung, C.; Ma, D.; Hong, S.; Kim, H. Optimization of strength and ductility of 2024 Al by equal channel angular pressing (ECAP) and post-ECAP aging. *Scr. Mater.* **2003**, *49*, 333–338.

11. Frint, P. Lokalisierungsphänomene Nach Kombinierter Hochgradig Plastischer Umformung Durch Extrusion und ECAP Einer 6000er-Aluminiumlegierung. Ph.D. Thesis, Chemnitz University of Technology, Chemnitz, Germany, 2015.

12. Lapovok, R.; Tóth, L.S.; Molinari, A.; Estrin, Y. Strain localisation patterns under equal-channel angular pressing. *J. Mech. Phys. Solids* **2009**, *57*, 122–136.

13. Zhilyaev, A.; Swisher, D.; Oh-Ishi, K.; Langdon, T.; McNelley, T. Microtexture and microstructure evolution during processing of pure aluminum by repetitive ECAP. *Mater. Sci. Eng. A* **2006**, *429*, 137–148.

14. Segal, V. Equal channel angular extrusion: from macromechanics to structure formation. *Mater. Sci. Eng. A* **1999**, *271*, 322–333.

15. Miyamoto, H.; Ikeda, T.; Uenoya, T.; Vinogradov, A.; Hashimoto, S. Reversible nature of shear bands in copper single crystals subjected to iterative shear of ECAP in forward and reverse directions. *Mater. Sci. Eng. A* **2011**, *528*, 2602–2609.

16. Prangnell, P.; Harris, C.; Roberts, S. Finite element modelling of equal channel angular extrusion. *Scr. Mater.* **1997**, *37*, 983–989.

17. Raab, G. Plastic flow at equal channel angular processing in parallel channels. *Mater. Sci. Eng. A* **2005**, *410*, 230–233.

18. Dumoulin, S.; Roven, H.; Werenskiold, J.; Valberg, H. Finite element modeling of equal channel angular pressing: Effect of material properties, friction and die geometry. *Mater. Sci. Eng. A* **2005**, *410*, 248–251.

19. Kim, H.S.; Seo, M.H.; Hong, S.I. Finite element analysis of equal channel angular pressing of strain rate sensitive metals. *J. Mater. Process. Technol.* **2002**, *130*, 497–503.

20. Yoon, S.C.; Kim, H.S. Finite element analysis of the effect of the inner corner angle in equal channel angular pressing. *Mater. Sci. Eng. A* **2008**, *490*, 438–444.

21. Kim, H.S.; Seo, M.H.; Hong, S.I. Plastic deformation analysis of metals during equal channel angular pressing. *J. Mater. Process. Technol.* **2001**, *113*, 622–626.

22. Park, J.W.; Suh, J.Y. Effect of die shape on the deformation behavior in equal-channel angular pressing. *Metall. Mater. Trans. A* **2001**, *32*, 3007–3014.

23. Li, S.; Bourke, M.; Beyerlein, I.; Alexander, D.; Clausen, B. Finite element analysis of the plastic deformation zone and working load in equal channel angular extrusion. *Mater. Sci. Eng. A* **2004**, *382*, 217–236.

24. Figueiredo, R.B.; Cetlin, P.R.; Langdon, T.G. Stable and unstable flow in materials processed by equal-channel angular pressing with an emphasis on magnesium alloys. *Metall. Mater. Trans. A* **2010**, *41*, 778–786.

25. Semiatin, S.; Delo, D.; Shell, E. The effect of material properties and tooling design on deformation and fracture during equal channel angular extrusion. *Acta Mater.* **2000**, *48*, 1841–1851.

26. Oruganti, R.; Subramanian, P.; Marte, J.; Gigliotti, M.F.; Amancherla, S. Effect of friction, backpressure and strain rate sensitivity on material flow during equal channel angular extrusion. *Mater. Sci. Eng. A* **2005**, *406*, 102–109.

27. Bowen, J.; Gholinia, A.; Roberts, S.; Prangnell, P. Analysis of the billet deformation behaviour in equal channel angular extrusion. *Mater. Sci. Eng. A* **2000**, *287*, 87–99.

28. Figueiredo, R.B.; Aguilar, M.T.P.; Cetlin, P.R. Finite element modelling of plastic instability during ECAP processing of flow-softening materials. *Mater. Sci. Eng. A* **2006**, *430*, 179–184.

29. Ghazani, M.S.; Vajd, A. Finite Element Simulation of Flow Localization During Equal Channel Angular Pressing. *Trans. Indian Inst. Met.* **2017**, *70*, 1323–1328.

30. Shutov, A.; Kreißig, R. Finite strain viscoplasticity with nonlinear kinematic hardening: Phenomenological modeling and time integration. *Comput. Methods Appl. Mech. Eng.* **2008**, *197*, 2015–2029.

31. Kießling, R.; Landgraf, R.; Scherzer, R.; Ihlemann, J. Introducing the concept of directly connected rheological elements by reviewing rheological models at large strains. *Int. J. Solids Struct.* **2016**, *97*, 650–667.

32. Frederick, C.O.; Armstrong, P. A mathematical representation of the multiaxial Bauschinger effect. *Mater. High Temp.* **2014**, *24*, 1–26

33. Voce, E. The relationship between stress and strain for homogeneous deformation. *J. Inst. Met.* **1948**, *74*, 537–562.

34. Perzyna, P. The constitutive equations for rate sensitive plastic materials. *Q. Appl. Math.* **1963**, *20*, 321–332.

35. Horn, T. Simulation und FE-Analyse der Verformungslokalisierung bei der ECAP-Umformung. Master's Thesis, Chemnitz University of Technology, Chemnitz, Germany, 2016.

36. Shutov, A.V.; Kreißig, R. Geometric integrators for multiplicative viscoplasticity: Analysis of error accumulation. *Comput. Methods Appl. Mech. Eng.* **2010**, *199*, 700–711.

37. Shutov, A.V.; Kuprin, C.; Ihlemann, J.; Wagner, M.F.X.; Silbermann, C. Experimentelle Untersuchung und numerische Simulation des inkrementellen Umformverhaltens von Stahl 42CrMo4 Experimental investigation and numerical simulation of the incremental deformation of a 42CrMo4 steel. *Materialwissenschaft Werkstofftechnik* **2010**, *41*, 765–775.

38. Silbermann, C.B.; Shutov, A.V.; Ihlemann, J. On operator split technique for the time integration within finite strain viscoplasticity in explicit FEM. *PAMM* **2014**, *14*, 355–356.

39. Yamaguchi, D.; Horita, Z.; Nemoto, M.; Langdon, T.G. Significance of adiabatic heating in equal-channel angular pressing. *Scr. Mater.* **1999**, *41*, 791–796.

40. Frint, P.; Wagner, M.F.X.; Weber, S.; Seipp, S.; Frint, S.; Lampke, T. An experimental study on optimum lubrication for large-scale severe plastic deformation of aluminum-based alloys. *J. Mater. Process. Technol.* **2017**, *239*, 222–229.

41. Frint, S.; Hockauf, M.; Frint, P.; Wagner, M.F.X. Scaling up Segal's principle of Equal-Channel Angular Pressing. *Mater. Des.* **2016**, *97*, 502–511.

42. Horn, T.; Silbermann, C.; Ihlemann, J. FE-Simulation based analysis of residual stresses and strain localizations in ECAP processing. *PAMM* **2017**, *16*, in press.

43. Frint, P.; Hockauf, M.; Dietrich, D.; Halle, T.; Wagner, M.F.X.; Lampke, T. Influence of strain gradients on the grain refinement during industrial scale ECAP. *Materialwissenschaft Werkstofftechnik* **2011**, *42*, 680–685.

44. Frint, P.; Hockauf, M.; Halle, T.; Wagner, M.F.X.; Lampke, T. The role of backpressure during large scale Equal-Channel Angular Pressing. *Materialwissenschaft Werkstofftechnik* **2012**, *43*, 668–672.

45. Frint, P.; Hockauf, M.; Halle, T.; Strehl, G.; Lampke, T.; Wagner, M.F.X. Microstructural Features and Mechanical Properties after Industrial Scale ECAP of an Al 6060 Alloy. *Mater. Sci. Forum* **2011**, *667*, 1153–1158.

46. Lefstad, M.; Pedersen, K.; Dumoulin, S. Up-scaled equal channel angular pressing of AA6060 and subsequent mechanical properties. *Mater. Sci. Eng. A* **2012**, *535*, 235–240.

47. Chaudhury, P.K.; Cherukuri, B.; Srinivasan, R. Scaling up of equal-channel angular pressing and its effect on mechanical properties, microstructure, and hot workability of AA 6061. *Mater. Sci. Eng. A* **2005**, *410*, 316–318.

48. Berndt, N.; Frint, P.; Böhme, M.; Wagner, M.F.X. Microstructure and mechanical properties of an AA6060 aluminum alloy after cold and warm extrusion. *Mater. Sci. Eng. A* **2017**, *707*, 717–724.

49. Beyerlein, I.J.; Tomé, C.N. Analytical modeling of material flow in equal channel angular extrusion (ECAE). *Mater. Sci. Eng. A* **2004**, *380*, 171–190.

50. Ebeling, W.; Feistel, R. *Chaos und Kosmos: Prinzipien der Evolution*; Spektrum Akad. Verlag: Heidelberg, Germany, 1994; pp. 109–118.

51. Ebeling, W. *Chaos, Ordnung und Information: Selbstorganisation in Natur und Technik*; Urania-Verlag: Leipzig, Germany, 1989; Volume 74, p. 40.

metals

MDPI

Article

On the Effect of Natural Aging Prior to Low Temperature ECAP of a High-Strength Aluminum Alloy

Sebastian Fritsch * and Martin Franz-Xaver Wagner

Chair of Materials Science, Institute of Materials Science and Engineering, Technische Universität Chemnitz, 09125 Chemnitz, Germany; martin.wagner@mb.tu-chemnitz.de
* Correspondence: sebastian.fritsch@mb.tu-chemnitz.de; Tel.: +49-371-531-37441

Received: 21 December 2017; Accepted: 16 January 2018; Published: 18 January 2018

Abstract: Severe plastic deformation (SPD) can be used to generate ultra-fine grained microstructures and thus to increase the strength of many materials. Unfortunately, high strength aluminum alloys are generally hard to deform, which puts severe limits on the feasibility of conventional SPD methods. In this study, we use low temperature equal-channel angular pressing (ECAP) to deform an AA7075 alloy. We perform ECAP in a custom-built, cooled ECAP-tool with an internal angle of 90° at −60 °C and with an applied backpressure. In previous studies, high-strength age hardening aluminum alloys were deformed in a solid solution heat treated condition to improve the mechanical properties in combination with subsequent (post-ECAP) aging. In the present study, we systematically vary the initial microstructure—i.e., the material condition prior to low temperature ECAP—by (pre-ECAP) natural aging. The key result of the present study is that precipitates introduced prior to ECAP speed up grain refinement during ECAP. Longer aging times lead to accelerated microstructural evolution, to increasing strength, and to a transition in fracture behavior after a single pass of low temperature ECAP. These results demonstrate the potential of these thermo-mechanical treatments to produce improved properties of high-strength aluminum alloys.

Keywords: equal-channel angular pressing (ECAP); low temperature; cryogenic deformation; SPD-processes; high strength aluminum alloy; AA7075; AlZnMgCu-alloy; ultra-fine grained (UFG)

---

1. Introduction

Severe plastic deformation (SPD) methods allow the generation of ultra-fine grained (UFG) microstructures [1,2]; equal-channel angular pressing (ECAP) is one of the most successful SPD approaches to produce high strength combined with good ductility in various metals and alloys [3–7]. The increase of strength is due to grain-refinement processes into the sub-micrometer range. In regions of high dislocation densities, low-angle grain boundaries are formed and then further transformed into high-angle grain boundaries, which leads to a reduction of grain size by cold work [8–10]. ECAP processing has been used to increase the strength of various age hardening aluminum alloys. Typically, these alloys are deformed in the solution heat treated condition at room temperature (RT), mainly for two reasons: (i) to increase the workability during the SPD process and (ii) to form homogeneously distributed precipitates in a subsequent (i.e., post-ECAP) heat treatment to further improve the mechanical properties of the material. However, there are some difficulties with ECAP processing of high-strength aluminum alloys such as AA7075; in previous studies, we discussed why ECAP at RT of this material leads to the formation of shear bands and to subsequent cracking [11,12]. Increasing the workability by increasing the deformation temperature up to about 100 °C does in principle allow a successful production of homogenous billets, but the mechanical properties are hardly improved because of increased recovery processes [13].

There is, however, a viable approach to increase the strength of hard to deform face-centered cubic (fcc) metals based on the insights of physical metallurgy: to considerably increase the dislocation density during deformation (and thus to benefit from grain size reduction), dynamic recovery processes need to be suppressed during the deformation process [14,15]. This can be achieved by performing SPD at low (or even cryogenic) temperatures, where the strain hardening capability of fcc metals is increased and strain localization and crack formation can be suppressed. Additionally, cross slip and dislocation annihilation processes are prevented because of the limited mobility of defects at low temperatures. This leads to an improved ductility and workability, specifically because the strain hardening stages III and IV are considerably extended [11,12]. The most well-known process that exploits the advantages of low temperature deformation is cryogenic rolling. This is a relatively simple procedure which can be performed in conventional roller mills. To ensure a limited increase of temperature because of plastic work, the materials typically need to be deformed in small incremental steps. They are typically cooled in liquid nitrogen, then placed into the rolling mill and subsequently deformed–typically by true plastic strains of about 0.1 per rolling step. For SPD, these steps therefore have to be repeated several times [5,16–18]. Using this procedure, accumulated true plastic strains of about three have been achieved in high strength aluminum alloys [13,19], which resulted in a strong increase of initial flow stress and a somewhat increased ductility when cryogenic rolling was combined with a suitable post-rolling heat treatment. Rolling of course always leads to plate geometries, which is a limiting parameter for the use of a material in technological applications that require bulk billets. Multidirectional forging [20] and ECAP [8–10] at low temperatures allow for SPD processing of bulk materials under well-defined deformation conditions. Especially ECAP, which conserves the cross section of the billets, allows for careful control of the deformation velocity and facilitates application of an active backpressure, offers great potential to further develop low temperature SPD processing.

Another important aspect in ensuring that ECAP can be performed on high strength aluminum alloys is to reduce dynamic strain aging. This microstructural effect is the result of the interaction between dislocations and solute atoms. Dynamic strain aging can be observed as serrated flow in the corresponding stress-strain curves and leads to the formation of deformation bands. These bands are macroscopically visible on the surface of tensile specimens. The mechanism is well known as Portevin-Le Chatelier effect (PLC effect) and is related to thermally activated processes. The formation of PLC bands therefore strongly depends on temperature and strain rate. In summary, many aluminum alloys exhibit a better formability at low temperatures because of two effects: an increased strain-hardening rate and suppressed PLC effects. Previous studies have shown that $-60\ ^\circ$C is a suitable temperature for successful ECAP processing of high strength aluminum alloys [12]. This temperature increases the ductility of the material by about 60% compared to the ductility at RT and dramatically reduces the susceptibility for discontinuous flow.

Besides the deformation temperature, the formation of precipitates after the solution heat treatment influences the occurrence of PLC effects. These strengthening precipitates can already be formed at RT, depending on aging time. Because of the high amount of copper, the alloy AA7075 is prone to an early formation of Guinier-Preston (GP) zones even at RT. Already very short natural aging times (10 min) can strongly influence the onset of serrated flow; with increasing natural aging time, the first serrations are shifted to higher stress/strain values, which also affects subsequent crack initiation and failure of the specimen. After the formation of a sufficiently larger volume fraction of GP zones, strain localization can be entirely suppressed because the higher amount of GP zones leads to a decreased density of mobile point defects [12,21].

The microstructure of the AA7075 alloy is strongly influenced by natural aging after a solution heat treatment, already within very short times. In the present study, we consider this important aspect, and we discuss how the formation of precipitates prior to ECAP influences the workability at low deformation temperatures. Furthermore, we show how the microstructure of the aluminum alloy AA7075 is affected by the deformation process depending on the amount of precipitates formed prior to processing. Finally, we discuss how these microstructural changes influence the mechanical properties after low temperature ECAP.

2. Materials and Methods

The aluminum alloy AA7075 was obtained from KUMZ, (Kamensk Uralsky, Russia), as an extruded bar with a cross-section of 15×15 mm^2. The chemical composition was 5.9 Zn-2.3 Mg-1.3 Cu-0.2 Fe-0.1 Si wt %. The material was solutionized at 475 °C for 2.25 h and subsequently water-quenched to RT. Afterwards, the solutionized material was naturally aged between 1 min and 24 h at RT, followed by ECAP-processing (the internal angle was 90°, which corresponds to an equivalent plastic strain of 1.1) at −60 °C in a custom-built ECAP die for one pass. The pressing speed was 20 mm/min and a backpressure of 176 MPa was applied to suppress shear localization or failure by cracking. The billets were covered with the lubricant Aero Shell Grease 33MS (molybdenum disulphide) to minimize friction. Detailed information about the design and functionality of the setup for low temperature ECAP, and about the choice of deformation parameters is given in our previous publications [12,22].

After processing, the material was fully naturally aged. In order to characterize the natural aging of the material, hardness measurements (Brinell hardness 2.5/62.5) were conducted in the longitudinal plane of the billets after different aging times. Average hardness values and standard deviations were calculated from five indents. For further mechanical characterization, tensile tests using a Zwick/Roell 20 kN tensile testing machine (Ulm, Germany) were performed on ECAP-processed conditions with different aging times (between 1 min and 24 h) prior to ECAP. The miniaturized tensile specimens had a gauge length of 5 mm with a cross section of 2×2 mm^2 and were tested at RT with an initial strain rate of 10^{-3} s^{-1}. Strains were determined with an optical digital image correlation system (Aramis by GOM, version 6.3.1, Braunschweig, Germany). This technique measures the displacement fields of a speckle pattern on the specimen surface. The surface deformation field is recorded during the test and can subsequently be used to determine uniaxial strains. For statistics, five specimens of each condition were tested. Typical microstructural features of the different material conditions were analyzed by transmission electron microscopy (TEM) in a H8100 (Hitachi, Tokyo, Japan) operated at 200 kV. The TEM samples were cut parallel to the processing direction and were first twin jet electro-polished and then polished using an argon-ion-beam. Finally, the fracture surfaces were investigated in a field emission scanning electron microscope (NEON40 by Zeiss, Jena, Germany). The acceleration distance was 5 kV with an operating distance of 4 mm. Besides the detector for secondary and back-scattered electrons, an in-lens detector was used because it provides an enhanced contrast that improves the analysis of fracture surfaces.

3. Results and Discussion

There is a strong tendency for the formation of strength increased precipitates during natural aging in the solution heat treated condition of high strength age hardening aluminum alloys. This effect can be demonstrated with simple hardness measurements. Figure 1 shows the evolution of hardness (HBW) of AA7075 after solutionizing, followed by natural aging at RT, compared to aging at −30 °C. The initial hardness values are about 88 HBW. A few minutes of natural aging already lead to a slight increase of hardness. After two hours, the hardness reaches a value of about 96 HBW. With a further increase of aging time, the hardness steadily increases. After 24 h of natural aging, hardness reaches a value of about 135 HBW. This is an increase of approx. 50% compared to the initial hardness value. This strong increase of hardness is due to the early formation of GP zones which is promoted by the high amount of copper. In comparison, aging at −30 °C suppresses the formation of precipitates which leads to a constant hardness. Therefore, deformation at a low temperature provides a couple of advantages—the strain hardening capability is increased, strain localization and crack formation can be suppressed, and the influence of precipitates (formed by natural aging prior to low temperature ECAP) on microstructural changes can be systematically investigated.

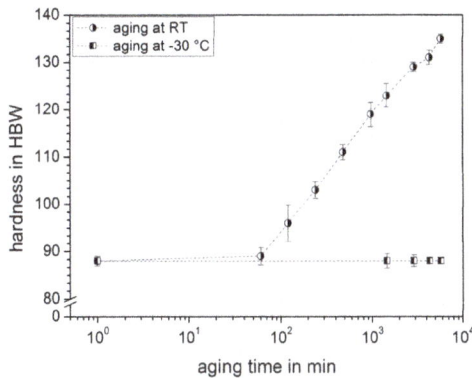

Figure 1. Evolution of hardness of the aluminum alloy AA7075 after solution heat treatment with subsequent aging at room temperature (RT) and at $-30\,^\circ$C. Lower temperatures suppress the formation of strength-increasing precipitates and leads to constant values of hardness compared to RT aging even after relatively long aging times.

In order to characterize the effect of the thermo-mechanical treatments studied here, we briefly discuss the correlation between processing parameters (specifically, the natural aging time prior to ECAP) and the corresponding microstructures. Figure 2 shows typical TEM bright field images of the conventional peak aged condition compared to the deformed material. The grain size of the conventional peak aged condition is about 4–7 µm (Figure 2a). As reported in [11] this condition shows nano-scaled precipitates in the matrix, with a size ranging from 5 to 15 nm and a platelet shape. The precipitates are homogenously distributed inside the grains. They become coarser and more sparsely distributed near the grain boundaries. In these regions, precipitation free zones with width of about 30 nm can be observed. The fine-grained microstructures after one ECAP pass at $-60\,^\circ$C are shown in Figure 2b–d. The large shear deformation immediately after solutionizing (this material condition is referred to as aged for 1 min throughout this paper, Figure 2b) results in the formation of characteristic shear bands with a width of approximately one micron. The dislocation density is considerably increased. This mechanism leads to the onset of recovery processes by formation of wall-like substructures. These sub-grain structures influence the final grain-size after SPD [23–26]. Characteristic regions in the TEM images can be identified as low angle grain-boundaries [27,28]. Adjacent to these dark areas with a high dislocation density, some regions appear much brighter, which corresponds to a much lower dislocation density. The microstructure exhibits a bimodal character. It is well known that the shear deformation during ECAP processing can be partitioned into regions with high and low deformation. These distinct regions are delimited by high-angle grain boundaries, which can already be formed locally after a single ECAP pass. As a result of natural aging (8 h, Figure 2c), strengthening precipitates (GP zones) are formed. Consequently, more dislocations are pinned during ECAP processing. This leads to a more pronounced formation of cell walls and sub-grain structures (Figure 2c). Therefore, the area fraction of less severely deformed regions decreases. This also leads to an increased number of high-angle grain boundaries, which can be observed in the severely deformed areas in Figure 2c. We note that the microstructure exhibits a similar grain-size distribution as material after ECAP processing with three passes without natural aging prior to the deformation, as reported in [22]. After natural aging for 24 h prior to ECAP, the dislocation density as well as the formation of sub-grain structures are further increased (Figure 2d). Regions with less shear deformation have almost disappeared. The microstructure is homogeneous, with an increased number of fine-grained regions compared to the conditions with shorter pre-ECAP aging times. This grain-size distribution after only one ECAP-pass is similar to a severely deformed condition after four passes, see also [22]. The TEM results summarized in Figure 2 clearly show that natural aging prior to ECAP can strongly influence the microstructural evolution. This effect is obviously related to the formation of GP zones, which leads to an increased number of stored and pinned dislocations, and which further results in increased dislocation densities and in the early formation of fine-grained microstructures.

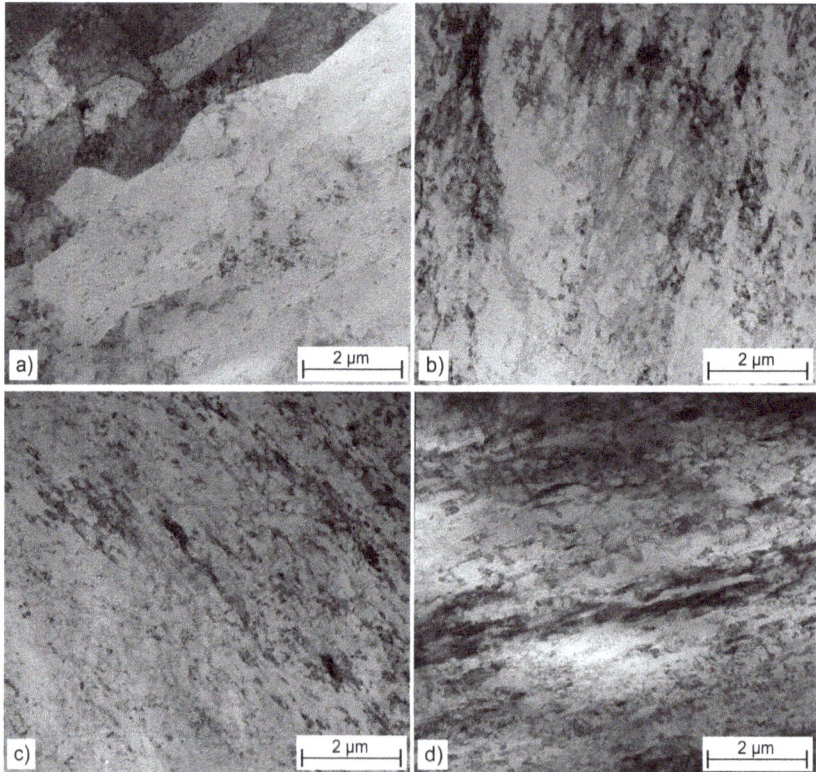

Figure 2. Transmission electron microscopy (TEM) bright field images of the AA7075 alloy in (a) conventional peak-aged condition with an average grain size of about 4–7 μm compared to the as-processed material after equal-channel angular pressing (ECAP). The material was deformed in the solutionized heat treated condition, with RT aging (prior to ECAP) for (**b**) 1 min; (**c**) 8 h and (**d**) 24 h. The increase of pre-ECAP natural aging time results in a homogeneous grain refinement with a distinct sub-grain microstructure.

The conventional concept of grain refinement as the result of the accumulation of dislocations and the formation of sub-grain structures and new grains is well known; the schematic illustration in Figure 3 extends this view to provide a broader picture of the relationship between grain refinement processes and precipitates formed prior to ECAP. Figure 3a represents a precipitation free, solid solution heat treated condition with some high-angle grain boundaries. Typically, SPD (like ECAP) processing of such material conditions allow generating (ultra)fine-grained microstructures, at first driven by the formation of subgrains. Natural aging prior to ECAP leads to the formation of GP zones (Figure 3b). Dislocations are pinned by these GP zones during the deformation. This leads to an increased dislocation density and a much more pronounced sub-grain network even after a single ECAP pass. With increasing natural aging times prior to deformation, the number of precipitates is further increased (Figure 3c), which provides additional points to pin dislocations, and again intensifies the formation of sub-grain structures and the grain refinement process.

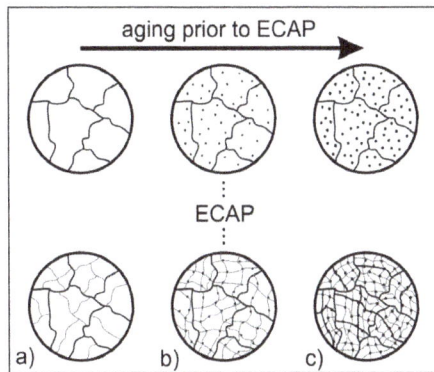

Figure 3. Schematic representation of the modified grain refinement process as a result of the formation of precipitates prior to severe plastic deformation (SPD) processing. Figures in the top row indicate the formation of precipitates as a function of increasing aging times. The images below show how the precipitates influence the grain refinement process during subsequent ECAP: (**a**) in the solution heat treated condition, the formation of sub-grain structures primarily results from dislocation interactions; (**b**) dislocations are additionally pinned by the precipitates formed before, which results in the formation of additional sub-grains; (**c**) longer aging times lead to an increased number of precipitates and further promotes the formation of dislocation networks and sub-grains after only one ECAP pass.

The modified microstructures of course have a strong influence on the mechanical properties of the material conditions that have been naturally aged prior to ECAP. In order to further characterize the effects of the thermomechanical treatments studied here, we briefly discuss the relation between microstructure and the corresponding mechanical properties. Figure 4 shows characteristic (engineering) stress-strain curves of the material conditions that were processed by ECAP and that were subjected to natural aging prior to the deformation for 1 min, 8 h and 24 h, respectively. An increased natural aging time results in a strong increase of strength while ductility is decreased. After performing a large number of tensile tests on different material conditions, characteristic mechanical parameters were determined. They are shown as a function of natural aging time prior to ECAP in Figure 5 (Figure 5a: strength values; Figure 5b, strain values). These data allow evaluating the aging kinetics. Tensile testing after solutionizing (labeled again as an aging time of 1 min) results in a yield strength (YS) of about 500 MPa (Figure 5a). YS remains almost constant for short pre-aging times up to about 2 h. The ultimate tensile strength (UTS) shows a similar behavior, with values of about 580 MPa. With increasing natural aging times, the strength increases continuously. After 720 min the yield strength and ultimate tensile strength values reach saturation values of about 550 and 630 MPa, respectively. This corresponds to an increase of about 10% compared to the conventional peak aged condition—we highlight that this is a noteworthy improvement since no post-ECAP heat treatment (which is typically used to further modify mechanical behavior) was used. The ductility is almost constant up to natural (pre-ECAP) aging times of 2 h (Figure 5b). Compared to the conventional peak aged condition, uniform elongation (UE) and elongation to failure (EF) values are slightly higher with increases of 9% and 12%, respectively. Further increased aging times lead to a pronounced decrease of ductility. Uniform elongation values drop rapidly with increasing pre-ECAP aging times. After 24 h of natural aging prior to ECAP, elongation to failure is less than 4%, so that necking could not even be observed. In summary, the microstructural changes associated with natural aging, specifically the nucleation of GP zones, clearly affect the mechanical properties of the material conditions studied here. An increasing volume fraction of GP zones during the natural aging process fully rationalizes the observed increases of hardness and strength (Figures 1 and 5a) as well as the reduction of ductility (Figure 5b).

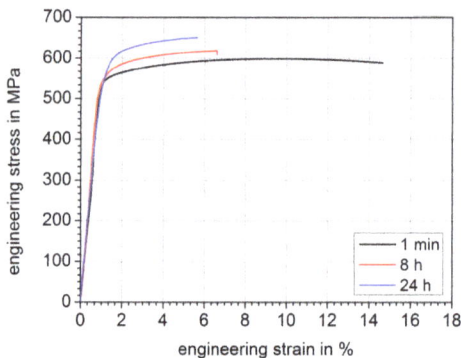

Figure 4. Characteristic engineering stress-engineering strain curves of AA7075 under quasi static loading at RT. Samples were naturally aged for 1 min, 8 h or 24 h prior to ECAP. Increasing the natural aging time prior to deformation results in increased strength and reduced ductility after one ECAP pass.

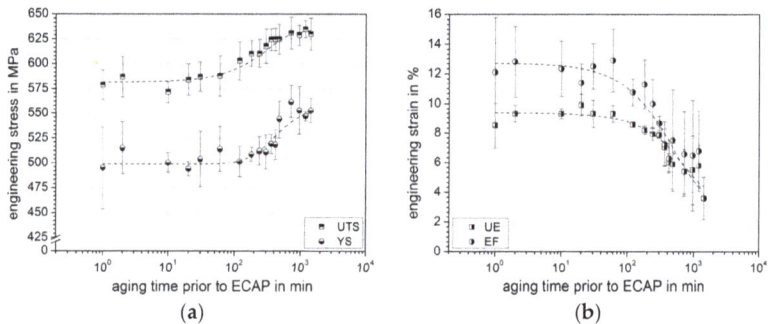

Figure 5. Characteristic values of (**a**) engineering stress and (**b**) engineering strain depending on aging at RT prior to ECAP in the solution heat treated condition for one pass at −60 °C. The increase of natural aging time prior to ECAP results in an increase of ultimate tensile strength (UTS) and yield strength (YS), while the ductility (uniform elongation (UE) and elongation to failure (EF)) is decreased.

The changes in terms of microstructural features and of the mechanical properties can also be directly related to an analysis of the corresponding fracture surfaces. The results of the SEM investigations are presented in Figure 6, with a focus on two representative material conditions: ECAP directly after solutionizing (Figure 6a,c,e) versus pre-ECAP aging of 24 h aging at RT (Figure 6b,d,f), respectively. The boxes mark areas of selected for higher magnification. Macroscopically, the failure of both tensile specimens can be described as shear fracture at angles of about 45° with respect to the loading direction. However, microscopically, there are differences between the two material conditions. The condition with only 1 min of natural aging predominantly shows smooth regions at low magnification (Figure 6a). These are zones of lower deformation (see also Figure 2b) and are enclosed by larger dimples (Figure 6c). The regions of ductile fracture or cavities are predominantly formed at grain- or sub-grain boundaries and have a width of about 1 μm. Plastic straining during tensile testing widens the pores, which then leads to the initiation of cracks. This results in local necking at the cavities and suppresses further deformation. Finally, the remaining regions between pores are ruptured by the formation of shear lips. Most of the fracture surfaces, however, are characterized by smooth and less pronounced dimples (figure 6e). These regions considerably contribute to the macroscopically stable deformation and result in higher ductility, Figure 5b. Natural aging for 24 h prior to ECAP results in the formation of GP zones, which leads to a much more reduced (sub-)grain size. The increased numbers of precipitates, grain- and sub-grain boundaries is associated with a significantly more pronounced formation of dimples on the fracture surface, which therefore appears much rougher even at low magnifications (Figure 6b). Higher resolutions confirm reduced dimple sizes (Figure 6d) compared to Figure 6c. Regions with smooth surfaces are hardly observed. Throughout the entire fracture surface, the macroscopic deformation takes place by the formation of honeycomb structures. These dimples have a size range of about 50 to 100 nm (Figure 6f) with a strongly deformed topography and very fine intervoid ligaments. With increasing plastic strain during tensile testing, the deformation is concentrated in these regions. Due to the low supporting cross-section, the material tends to rapid local necking, which decreases macroscopic ductility. Finally, we note that the very small precipitates (in the size range of about 20 nm) often act as internal notches and thus also lead to a further decrease of ductility (Figure 6f).

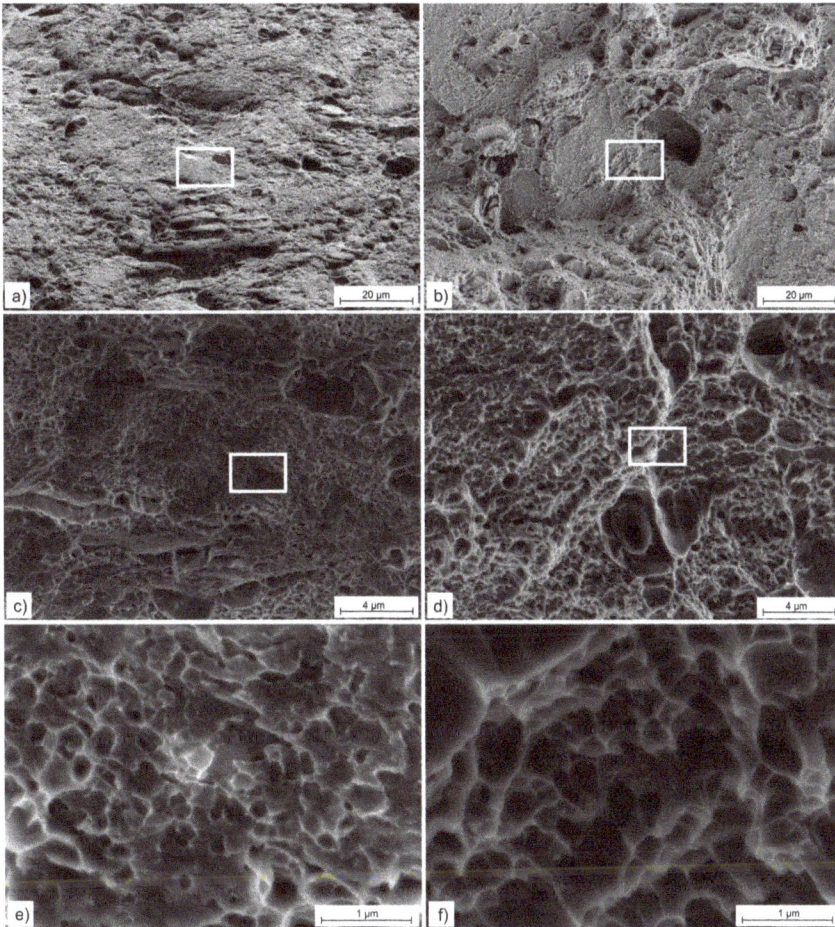

Figure 6. SEM micrographs of the fracture surfaces of AA7075 after one ECAP pass under cryogenic conditions with pre-ECAP natural aging of (**a,c,e**) 1 min and (**b,d,f**) 24 h. Boxes mark the areas studied further at higher magnification. Natural aging prior to the deformation process results in a transition from a mixed fracture surface with smooth shear parts and dimple structures to fracture surfaces with very fine dimples.

4. Summary and Conclusions

We have studied the influence of natural aging of a solid solution heat treated AA7075 alloy (prior to SPD) on the microstructure and the mechanical properties after equal-channel angular pressing. Because of an increased strain hardening capability of this alloy, low deformation temperatures result in an improved workability and increased ductility. Furthermore, low temperatures suppress the formation of precipitates. Therefore, we deformed the material at $-60\ ^\circ$C by ECAP and varied natural aging times prior to the deformation. We systematically investigated how the formation of strength-increasing precipitates influences the microstructure after SPD. As documented by transmission electron microscopy, low temperature ECAP combined with natural aging prior to the deformation results in relevant microstructural changes, i.e., grain refinement in the sub-micrometer range. A short pre-aging time results in the formation of a bimodal grain structure, where regions with fine-grained structures as well as regions with less deformation can be observed. When the aging time prior to deformation is increased, the number of precipitates is strongly increased. This leads to a more homogeneous microstructure with finer sub-structures. The documented microstructural changes are directly related to the post-ECAP mechanical properties that were further characterized by tensile testing. A reduced grain size leads to an increase of strength and to a reduced ductility and also affects fracture behavior.

Acknowledgments: The authors gratefully acknowledge funding by the German Research Foundation (Deutsche Forschungsgemeinschaft, DFG) within the framework of SFB 692 (project A1).

Author Contributions: Sebastian Fritsch and Martin Franz-Xaver Wagner conceived of this study. Sebastian Fritsch performed and analyzed the experiments, prepared the figures, and drafted the manuscript. Martin Franz-Xaver Wagner discussed the results and analysis and helped writing the manuscript.

Conflicts of Interest: The authors declare no conflict of interest.

References

1. Zhilyaev, A.; Langdon, T.G. Using high-pressure torsion for metal processing: Fundamentals and applications. *Prog. Mater. Sci.* **2008**, *53*, 893–979. [CrossRef]
2. Frint, S.; Hockauf, M.; Frint, P.; Wagner, M.F.-X. Scaling up Segal's principle of Equal-Channel Angular Pressing. *Mater. Des.* **2016**, *97*, 502–511. [CrossRef]
3. Valiev, R.Z.; Langdon, T.G. Principles of equal-channel angular pressing as a processing tool for grain refinement. *Prog. Mater. Sci.* **2006**, *51*, 881–981. [CrossRef]
4. Frint, P.; Hockauf, M.; Halle, T.; Strehl, G.; Lampke, T.; Wagner, M.F.-X. Microstructural features and mechanical properties after industrial scale ECAP of an Al-6060 alloy. *Mater. Sci. Forum* **2010**, *667*, 1153–1158. [CrossRef]
5. Zhao, Y.H.; Liao, X.Z.; Cheng, S.; Ma, E.; Zhu, Y.T. Simultaneously increasing the ductility and strength of nanostructured alloys. *Adv. Mater.* **2006**, *18*, 2280–2283. [CrossRef]
6. Valiev, R.Z.; Langdon, T.G. Developments in the use of ECAP processing for grain refinement. *Rev. Adv. Mater. Sci.* **2006**, *13*, 15–26.
7. Segal, V.M.; Reznikov, A.E.; Drobyshevskiy, A.E.; Kopylov, V.I. Plastic working of metals by simple shear. *Russ. Metall.* **1981**, *1*, 971–974.
8. Ferrasse, S.; Segal, V.M.; Alford, F.; Kardokus, J.; Strothers, S. Scale up and application of equal-channel angular extrusion for the electronics and aerospace industries. *Mater. Sci. Eng. A* **2008**, *493*, 130–140. [CrossRef]
9. Kim, W.J.; Kim, J.K.; Kim, H.K.; Park, J.W.; Jeong, Y.H. Effect of post equal-channel-angulat-pressing agingon the modified 7075 Al alloy containing Sc. *J. Alloys Compd.* **2008**, *450*, 222–228. [CrossRef]
10. Zhao, Y.H.; Liao, X.Z.; Jin, Z.; Valiev, R.Z.; Zhu, Y.T. Microstructures and mechanical properties of ultrafine grained 7075 Al alloy processed by ECAP and their evolutions during annealing. *Acta Mater.* **2004**, *52*, 4589–4599. [CrossRef]

11. Fritsch, S.; Hunger, S.; Scholze, M.; Hockauf, M.; Wagner, M.F.-X. Optimisation of thermo mechanical treatments using cryogenic rolling and aging of the high strength aluminium alloy AlZn5.5MgCu (AA7075). *Materialwiss. Werkstofftech.* **2011**, *42*, 573–579. [CrossRef]

12. Fritsch, S.; Scholze, M.; Wagner, M.F.-X. Influence of thermally activated processes on the deformation behavior during low temperature ECAP. *IOP Conf. Ser. Mater. Sci.* **2016**, *118*, 1–13. [CrossRef]

13. Fritsch, S.; Hockauf, M.; Schönherr, R.; Hunger, S.; Meyer, L.W.; Wagner, M.F.-X. Investigation of the influence of ECAP and cryogenic rolling on the mechanical properties of the aluminium alloy 7075. *Materialwiss. Werkstofftech.* **2010**, *41*, 697–703. [CrossRef]

14. Seeger, A.; Haasen, P. Density changes of crystals containing dislocations. *Philos. Mag.* **1958**, *3*, 470–485. [CrossRef]

15. Nabarro, F.R.N. Work hardening and dynamical recovery of FCC metals in multiple glide. *Acta Metall.* **1989**, *37*, 1521–1546. [CrossRef]

16. Magalhães, D.C.C.; Hupalo, M.F.; Cintho, O.M. Natural aging behavior of AA7050 Al alloy after cryogenic rolling. *Mater. Sci. Eng. A* **2014**, *593*, 1–7. [CrossRef]

17. Panigrahi, S.K.; Jayaganthan, R. Development of ultrafine grained high strength age hardenable Al 7075 alloy by cryorolling. *Mater. Des.* **2011**, *32*, 3150–3160. [CrossRef]

18. Deschamps, A.; Livet, F.; Bréchet, Y. Influence of predeformation on ageing in an Al-Zn-Mg alloy-I. Microstructure evolution and mechanical properties. *Acta Mater.* **1998**, *47*, 281–292. [CrossRef]

19. Panigrahi, S.K.; Jayaganthan, R. Effect of annealing on thermal stability, precipitate evolution, and mechanical properties of cryorolled Al 7075 alloy. *Metall. Mater. Trans. A* **2011**, *42*, 3208–3217. [CrossRef]

20. Rao, P.N.; Singh, D.; Jayaganthan, R. Mechanical properties and microstructural evolution of Al 6061 alloy processed by multidirectional forging at liquid nitrogen temperature. *Mater. Des.* **2014**, *56*, 97–104. [CrossRef]

21. Leacock, A.G.; Howe, C.; Brown, D.; Lademo, O.G.; Deering, A. Evolution of mechanical properties in a 7075 Al-alloy subject to natural ageing. *Mater. Des.* **2013**, *49*, 160–167. [CrossRef]

22. Fritsch, S.; Scholze, M.; Wagner, M.F.-X. Cryogenic forming of AA7075 by equal-channel angular pressing. *Materialwiss. Werkstofftech.* **2012**, *43*, 561–566. [CrossRef]

23. Kocks, U.F. Laws for work-hardening and low-temperature creep. *J. Eng. Mater. Technol. ASME* **1976**, *98*, 76–85. [CrossRef]

24. Iwahashi, Y.; Furukawa, M.; Horita, Z.; Nemoto, M.; Langdon, T.G. Microstructural characteristics of ultrafine-grained aluminum produced using equal-channel angular pressing. *Metall. Mater. Trans. A* **1998**, *29*, 2245–2252. [CrossRef]

25. Mughrabi, H. Dislocation wall and cell structures and long-range internal stresses in deformed metal crystals. *Acta Metall.* **1983**, *31*, 1367–1379. [CrossRef]

26. Xu, C.; Furukawa, M.; Horita, Z.; Langdon, T.G. The evolution of homogeneity and grain refinement during equal-channel angular pressing: A model for grain refinement in ECAP. *Mater. Sci. Eng. A* **2005**, *398*, 66–76. [CrossRef]

27. Dalla Torre, F.; Van Swygenhoven, H.; Victoria, M. Nanocrystalline electrodeposited Ni: Microstructure and tensile properties. *Acta Mater.* **2002**, *50*, 3957–3970. [CrossRef]

28. Wang, N.; Wang, Z.; Aust, K.T.; Erb, U. Isokinetic analysis of nanocrystalline nickel electrodeposits upon annealing. *Acta Mater.* **1997**, *45*, 1655–1669. [CrossRef]

metals

MDPI

Article

Microstructural Evolution during Severe Plastic Deformation by Gradation Extrusion

Philipp Frint [1],*, Markus Härtel [1], René Selbmann [2], Dagmar Dietrich [1], Markus Bergmann [2], Thomas Lampke [1], Dirk Landgrebe [2] and Martin F.X. Wagner [1]

[1]   Institute of Materials Science and Engineering, Chemnitz University of Technology, D-09107 Chemnitz, Germany; markus.haertel@mb.tu-chemnitz.de (M.H.); dagmar.dietrich@mb.tu-chemnitz.de (D.D.); thomas.lampke@mb.tu-chemnitz.de (T.L.); martin.wagner@mb.tu-chemnitz.de (M.F.X.W.)

[2]   Fraunhofer Institute for Machine Tools and Forming Technology, D-09126 Chemnitz, Germany; rene.selbmann@iwu.fraunhofer.de (R.S.); markus.bergmann@iwu.fraunhofer.de (M.B.); dirk.landgrebe@iwu.fraunhofer.de (D.L.)

*   Correspondence: philipp.frint@mb.tu-chemnitz.de; Tel.: +49-371-531-38218

Received: 21 December 2017; Accepted: 24 January 2018; Published: 27 January 2018

Abstract: In this contribution, we study the microstructural evolution of an age-hardenable AA6082 aluminum alloy during severe plastic deformation by gradation extrusion. A novel die design allowing an interruption of processing and nondestructive billet removal was developed. A systematic study on the microstructure gradient at different points of a single billet could be performed with the help of this die. Distinct gradients were investigated using microhardness measurements and electron microscopy. Our results highlight that gradation extrusion is a powerful method to produce graded materials with partially ultrafine-grained microstructures. From the point of view of obtaining an ultrafine-grained surface layer with maximum hardness, only a small number of forming elements is needed. It was also found that large incremental deformation by too many forming elements may result in locally heterogeneous microstructures and failure near the billet surface caused by localization of deformation. Furthermore, considering economical aspects of processing, fewer forming elements are preferred since several processing parameter-related cost factors are then significantly lower.

Keywords: severe plastic deformation (SPD); microstructural gradient; ultrafine-grained (UFG); gradation extrusion; aluminum alloy; grain refinement

1. Introduction

Severe Plastic Deformation (SPD) processes are efficient techniques to increase a material's strength by strain hardening and grain refinement [1,2]. Many studies have been focused on the most prominent SPD processes like equal-channel angular pressing (ECAP) [3–6], accumulative roll-bonding (ARB) [7], and high-pressure torsion (HPT) [8,9]. One of the main goals of most SPD processes is the introduction of large strains into the billet, typically with the aim of a homogeneous distribution [10–13]. In order to reduce microstructural gradients and to homogenize the billet's properties, many SPD methods make use of repeated deformation steps [14–17] until a certain homogenization of the microstructure, as well as the corresponding mechanical properties, is achieved. Multiple passes of ECAP, for example, are often used to produce homogeneous bulk ultrafine-grained materials [18,19]. Much research has been performed in order to find optimum processing routes for multi-pass ECAP with enhanced grain refinement rates and maximum homogeneity [20–22].

In contrast, for some practical applications there is a need for materials that provide a distinct gradient of mechanical properties. Classical approaches to obtaining property/microstructure gradients are, e.g., shot peening of ductile materials and heat treatment methods like the case hardening of steels. There are, furthermore, many techniques of surface coating which are applicable for almost every material, but are often accompanied by the formation of substrate-coating interfaces that can be disadvantageous for certain applications. For tailoring specific gradient microstructures and mechanical properties, SPD methods have rarely been used [23–26]. In the present work, a distinct microstructural gradient is created by a heterogeneous introduction of severe plastic deformation. Based on the principle of direct extrusion processing, a novel approach of gradation extrusion has recently been developed [27,28]. This bulk forming method is basically a type of direct impact extrusion that has also been registered for patent [29]. Using several geometrical cavities in the die that act as forming elements as well as enforcing a final reduction of the diameter of the rod-shaped billets (see Figure 1), a well-defined graded microstructure with a severely deformed surface region can be generated. The forming elements are comparatively small with regards to the overall die size. They can be varied in number, size, and arrangement and they generate the specific conditions needed for severe plastic deformation. Depending on the tool design, a specific local gradient in terms of grain size near the mantle surface can be produced [30]. The forming technique requires sufficient backpressure to completely fill the cavities with the material between the forming elements. This backpressure is generated by the final diameter reduction. Another important requirement for effective grain refinement is a large induced equivalent strain. The gradation extrusion principle is therefore characterized by a stepwise generation of local plastic strain at the forming elements which is accumulated during the deformation path through the corresponding tool [28,29].

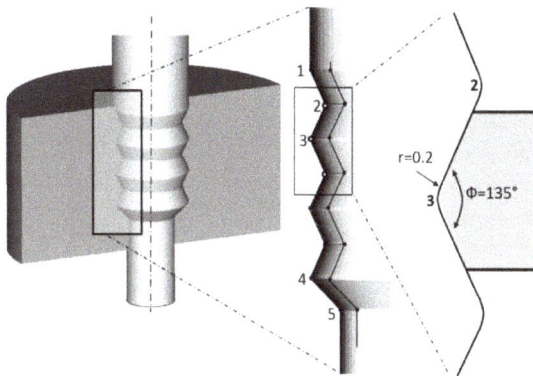

Figure 1. Gradation extrusion principle: Forming elements with undercuts, having an angle of 135° and a corner radius of 0.2 mm.

This approach has a certain potential to produce an ultrafine-grained microstructure which can extend up to a depth of several millimeters from the billet's surface. This layer is characterized by an increased strength while the inner material exhibits a high ductility due to its coarse-grained microstructure. Since the combination of local high strength and good ductility has high relevance for many practical applications, the investigated technological approach of gradation extrusion seems promising for bridging the gap between SPD methods refining the complete volume and conventional methods that only create surface strengthening in the micrometer regime. For an efficient tailoring of graded materials that are perfectly fit for a later application, it is important to generate an in-depth scientific understanding of microstructural evolution during processing and of the resulting (local) microstructure–properties relationships.

2. Materials and Methods

In order to generate graded rod-shaped materials, the process of gradation extrusion was developed. For detailed characterization of the microstructural evolution during processing, a unique extrusion die was designed and manufactured.

In contrast to conventional extrusion dies, the die is separated into an upper part and a lower die part. The separation plane is in the longitudinal direction and contains the axis of symmetry. This design allows for the removal of the processed billet. This is a basic requirement for a systematic characterization of microstructural evolution since the steady-state material flow is stopped at a certain point of extrusion. The billet can easily be removed without any damage or additional influence on the microstructure. The die parts are closed and compressed with a hydraulic press during forming, as presented in greater detail in [27]. The second distinct feature is another separation of the die into a gradation element and an extrusion element (see Figure 2). This design also allows different combinations (i.e., different relative contributions to the overall deformation) of gradation and extrusion.

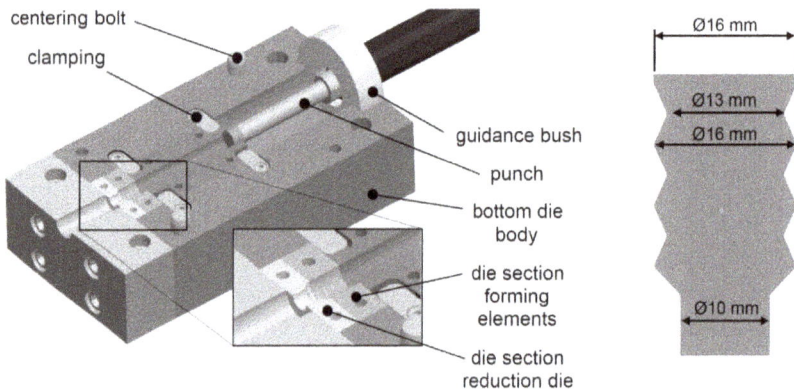

Figure 2. Tool system design for experiments [27] and channel diameters at the forming elements.

The die was mounted in a hydraulic press with a maximum capacity of 15 MN in order to close the die parts. An additional horizontal hydraulic cylinder with a maximum capacity of 2.5 MN provided the necessary pressing force for gradation extrusion. The gradation extrusion was performed at room temperature. Due to the low pressing speed of 1 mm/s during the extrusion experiments, quasi-adiabatic heating was prevented. Note that the pressing speed is a major processing factor that influences several aspects of the forming process like temperature gradients and failure mechanisms. Further details on the influence of pressing speed in conjunction with other parameters such as die geometry are discussed elsewhere [28]. In order to reduce friction during the gradation extrusion experiments, a commercial lubricant spray containing molybdenum disulfide and graphite that is commonly used for SPD processes [31] was applied on the cavity surface. The initial diameter of the billets was 16 mm and the minimum diameter of the concave forming elements was 13 mm. The maximum diameter of the convex forming elements was 16 mm. The final reduction leads to a diameter of 10 mm (see Figure 2).

In this study, gradation extrusion was performed using an age-hardenable aluminum alloy AA6082. The chemical composition of the investigated material can be found in Table 1. Prior to processing, the initial AA6082 alloy was solid solution annealed and naturally aged (T4). This alloy is often used for extrusion due to its high formability and strength. The initial material was commercially hot extruded and exhibits a typical near-surface zone characterized by large equiaxed grains, which results from dynamic recrystallization and pronounced grain growth in this highly strained region [32–34].

Table 1. Chemical composition of the investigated AA6082 alloy.

Element	Mg	Si	Fe	Mn	Cu	Cr	Zn	Ti	Al
wt %	0.83	0.9	0.17	0.56	0.047	0.008	0.005	0.019	bal.

After interrupted gradation extrusion, the billet was removed from the die and split lengthwise for mechanical and microstructural investigations. Hardness mapping (HV0.5) of an extruded billet was performed with the help of an automatic hardness tester (KB250BVRZ, KB Prüftechnik GmbH, Hochdorf-Assenheim, Germany). An equidistant grid was prepared with a spacing of 0.5 mm between individual indents and at a distance of 0.2 mm from the billet's outer surface.

Microstructural evolution during gradation extrusion was investigated using the opposite part of the lengthwise-split billet. Surfaces of longitudinal sections were prepared as described in [35] with a final finishing step of vibrational polishing. All micrographs presented in this paper were recorded with a viewing direction parallel to the normal direction of the specimen. The microstructure was investigated by scanning electron microscopy (SEM) using a field-emission device (Zeiss NEON 40 EsB, Zeiss, Jena, Germany) equipped with a backscatter electron detector (BSD) and an electron backscatter diffraction (EBSD) system (OIM 5.2, EDAX TSL, Mahwah, NJ, USA) which was operated at 15 kV with the 60 µm aperture in high current mode. EBSD data sets were typically recorded in regions of interest of 250 µm × 1000 µm with a step size of 3 µm, and of 15 µm × 15 µm with a step size of 50 nm, respectively. The EBSD data were subjected to a slight cleanup procedure comprising neighbor confidence index (CI) correlation and grain CI standardization with a minimum confidence index of 0.1. Parameters for grain size and grain aspect ratio determination were set to 15° tolerance angle, a minimum of 5 hits per grain, and a minimum CI of 0.1.

In order to compare the effective strains during gradation extrusion with other SPD processes, Finite Element (FE) simulations were performed. The finite element simulation includes work hardening of the material and friction between die and sample. The simulation setup and processing simulation were performed using the software Simufact Forming (version 13.3). The processes were set up as rotationally symmetric 2D models since the investigated die geometries are rotationally symmetric. A total of 7209 elements (element length 0.3 mm) of type advancing front quad were used. Both tool parts (punch and die) were defined as rigid bodies. The material and friction description used for the numerical simulation are summarized in Table 2 [28], where φ corresponds to true strain.

Table 2. Main parameters of the finite element (FE) model.

Material	$k_f = K \cdot \varphi^n$	$K = 400 \frac{N}{mm^2}; n = 0.094$
Friction	Coulomb, maximum shear stress (combined)	$\mu = 0.2; m = 0.08$

3. Results and Discussion

Figure 3 shows the distribution of effective (plastic) strain obtained from the FE simulation representing the steady state of material flow. A pronounced gradient of effective strain was found from the interior region to the surface. This gradient increases with every forming element in the axial direction. The final cross-sectional reduction leads to an additional introduction of strain, resulting in a final gradient ranging approximately from 1.4 (interior) to 6.9 (surface). The process basically fulfills the requirements for severe plastic deformation processes (SPD processes) since large effective strains, load path changes, high hydrostatic pressure, and temperatures below recrystallization temperature are applied. These results show that processing by gradation extrusion has a high potential for grain refinement by SPD, which leads to the assumption that distinct microstructural gradients may occur [28]. For a detailed analysis of the microstructure discussed below, five distinct locations are highlighted in Figure 3, representing interesting stages of deformation history during processing.

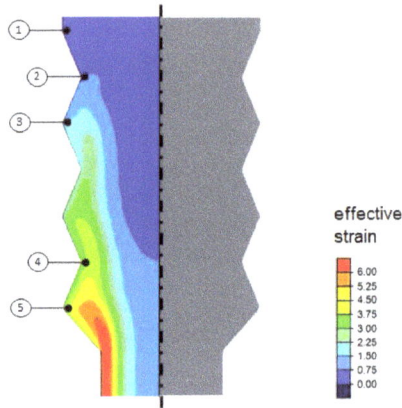

Figure 3. Distribution of effective strain determined from the FE simulation. Five distinct locations (1–5) for microstructural analysis, representing interesting stages of deformation history.

Figure 4 shows a processed billet of the investigated AA6082 alloy. The billet exhibits a smooth surface and no cracks have been found. It is also obvious that the cavities of all forming elements have been sufficiently filled with material, which indicates that the process of gradation extrusion was successfully performed. Both this experimental observation and the results of the FE simulation highlight that the requirements for a well-defined microstructural refinement by SPD are fulfilled.

Figure 4. AA6082 billet processed by interrupted gradation extrusion.

The initial material for gradation extrusion was first analyzed by SEM and EBSD (Figure 5). The microstructure of the commercially extruded AA6082 alloy is characterized by elongated fine grains in the interior region of the billet and by a surface area with a considerably coarsened grain structure. The mean grain size of the interior region is 10.2 (±2.3) μm and grains have a grain shape aspect ratio of 3. The shape of the grains in the interior region can be rationalized with the extrusion process of the initial material. This structure is a typical extrusion pancake structure [36]. The surface area has a thickness of approximately 200 μm and is characterized by almost equiaxed large grains with a diameter of about 150 μm. The exceptionally large shear strains introduced into the surface area, accompanied by quasi-adiabatic heating during commercial direct extrusion in conjunction with an overall high pressing temperature, lead to intensive dynamic recrystallization and grain growth in the surface area [36]. Furthermore, a typical strong <111> fiber texture accompanied by a <100> fiber texture caused by extrusion is found (see also Figure 5).

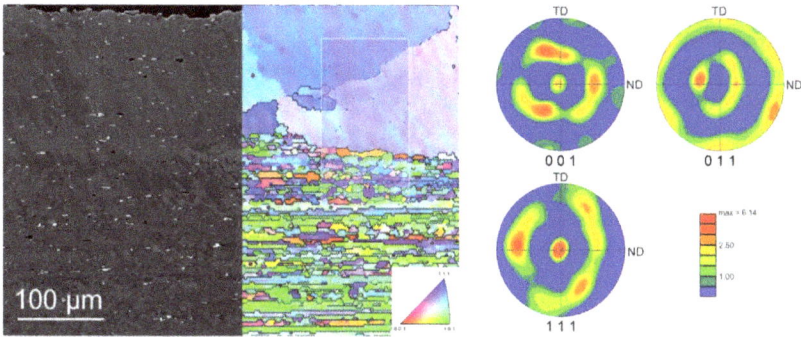

Figure 5. Microstructure of the initial material prior to gradation extrusion (left to right: SEM, electron backscatter diffraction (EBSD) map, pole figures calculated from EBSD data). TD: Transverse; ND: normal direction.

As already shown by the FE simulation (Figure 3), processing by gradation extrusion leads to microstructural changes due to incremental shear deformation by severe plastic deformation, especially in the surface area [36]. Figure 6 shows the contour of a gradation-extruded lengthwise-cut billet. The incremental deformation is realized by three concave and three convex, alternatingly arranged forming elements and by a subsequent reduction of the circular cross section. With the help of a hardness mapping on this lengthwise-split extrusion billet, a pronounced hardness gradient was documented. It was found that hardness does not change in the interior region while it rapidly increases (by 20%) in the surface area along the first concave and convex forming elements. At Point 3, the maximum hardness at the surface was observed. Moreover, these maximum hardness values were measured up to a depth of approximately 2 mm from the surface, which is the deepest uniformly distributed hardness all over the specimen.

From the point of view of obtaining a maximum gradation in terms of hardness, the processing should be finished after Point 3: Additional forming elements for gradation extrusion lead to a decrease in hardness and a less-pronounced gradient of mechanical properties at the second concave and convex forming elements. This can be rationalized with dynamic recovery and/or recrystallization due to undesired heating by friction and severe local straining. At the last concave and convex forming element (Point 4), an increase of hardness and, therefore, an increased gradient is observed again. It is assumed that a repeated increase in strain hardening caused by dislocation and defect multiplication of the dynamically recrystallized material leads to the observed increase in hardness. As a result, the last two concave and convex forming elements do not increase the hardness gradient. The final cavity involving a reduction of the cross section at Point 5 results in a homogenization of the hardness gradient, while the magnitude of the gradient itself remains. The zone of maximum hardness exhibits a constant width of approximately 1 mm and a thin transition region to the softer interior region of the billet is hardly affected by gradation extrusion.

The history of deformation during gradation extrusion that first has been characterized by hardness measurements was furthermore rationalized by accompanying microstructural investigations at the same five points of interest (Figure 6). For this investigation, the split billet was analyzed with the help of EBSD. Figure 6b–f corresponds to the points from the hardness mapping, respectively. At all five points, an EBSD measurement (at the same point) was performed using identical parameters. Each of Figure 6b–f shows an orientation map (OM) and the corresponding image quality map (IQ) below. This methodology allows a sufficient comparison of grain sizes, orientation data, and elongation of the grains at all five investigated points. The (severely strained) surface area is located on the left for each OM/IQ image. IQ maps give a qualitative indication of the magnitude of inherent strain. A brighter IQ image is associated with less lattice distortion and, therefore, lower inherent strains, and highly strained regions as well as grain boundaries typically show low/dark band contrast [37]. Note that microstructural features of the surface region are discussed in greater detail below (see also Figure 7).

Figure 6b corresponds to Point 1 and represents the initial condition of the material prior to gradation extrusion. As already shown in Figure 5, the surface area (~200 μm thickness) is characterized by almost equiaxed large grains (diameter ~150 μm) resulting from a pronounced dynamic recrystallization which is also confirmed by lower hardness values in this region. Underneath the coarse-grained surface region, the microstructure is characterized by elongated fine grains and a homogeneous pancake-like structure. The bright IQ image reveals a very low magnitude of inherent strain resulting from the commercial hot extrusion accompanied by pronounced recovery/recrystallization.

Figure 6c corresponds to Point 2 and represents the condition after the first concave forming element with an effective strain of approximately 1.4. Due to this severe plastic deformation, grain refinement occurs (comparable to other SPD microstructures) within the first 150 μm from the surface. The formerly coarse grains of the surface region are considerably refined compared to those of the interior region. Furthermore, a slight rotation of the grains can be observed originating from the material's flow through the cavity at Point 2. From the orientation map, it can be concluded that after about 300 μm from the surface, the microstructure is almost similar to that of the interior region of the initial material. However, the introduced work hardening indicated by the dark IQ map extends to about 500 μm. Work hardening and grain refinement during this local severe plastic deformation led to a considerable increase in hardness.

Figure 6d corresponds to Point 3 and represents the condition after the first concave and convex forming element with an effective strain of approximately 2.1. The maximum hardness at this point can be rationalized by the additional strain hardening and grain refinement induced by SPD. Considerable grain refinement is found within a region of about 250 μm from the billet surface. Again, the highly strained region is much larger with approximately 700 μm from the surface, as shown by the corresponding IQ map. Furthermore, a slight change of the dominating crystallographic orientations is found in the interior region. The pancake structure does not proceed parallel to the surface anymore (originating from the extruded initial material) but fits along the geometry of the cavity inside the gradation extrusion die. The passing of one concave and one convex forming element led to exceptional grain refinement and work hardening of the surface region, which results in the maximum hardness.

Figure 6e corresponds to Point 4 and represents the condition after three concave and three convex forming elements with an effective strain of approximately 3.6. The shape of the grains appears to be more equiaxed than at the previously investigated point. The globular grain shape is an indication of dynamic recrystallization between Points 3 and 4, driven by friction- and strain-induced (quasi-adiabatic) heating and severe lattice distortion from an overall high defect density. There are almost no traces of a pancake structure left. The bright IQ map confirms the assumption of pronounced dynamic recrystallization and recovery all over the entire area investigated by EBSD (250 μm × 1000 μm). This is additionally confirmed by the decreased hardness between Points 3 and 4 in the surface area. The high hardness at Point 4 is mainly attributed to the fine-grained microstructure and not to work hardening (bright IQ map).

Interestingly, some shear localization can be observed in the EBSD measurement, which might be an indication for a very low local strain hardening rate of the material in this condition. One possible reason might be the evolution of the texture during processing, which might consequently lead to some kind of softening or kinematic hardening that supports the localization of strain even while recrystallization takes place. Additional microstructural analysis is needed for a more detailed understanding of the relationship between certain microstructural features and the occurrence of shear localization. Such shear localizations may also induce material damage at the surface of the specimen, which supports the conclusion that the gradation process is finalized just after the first cavity and no additional forming elements are needed. This finding also supports the need for careful characterization of the material's shear behavior [38] from both experimental and theoretical points of view to minimize the risk of shear-induced cracking in SPD processes.

Figure 6f corresponds to Point 5 and represents the point after the final reduction of the specimen with an effective strain of approximately 4.3. This final reduction of the diameter leads to a reduction of the width of the maximum hardness region. This also results in an intensification of the hardness gradient (thin transition from surface to interior region) and in a homogenization of the microstructure within both regions.

Considering the results from Figure 6, it can be assumed that only one forming element and a final reduction cavity for the gradation extrusion process are sufficient from the point of gaining maximum hardness and maximum gradients. This is also beneficial from an economic point of view, since processing consumes less time and friction is simultaneously reduced, which decreases pressing forces and reduces the risk of material damage at the billet surface.

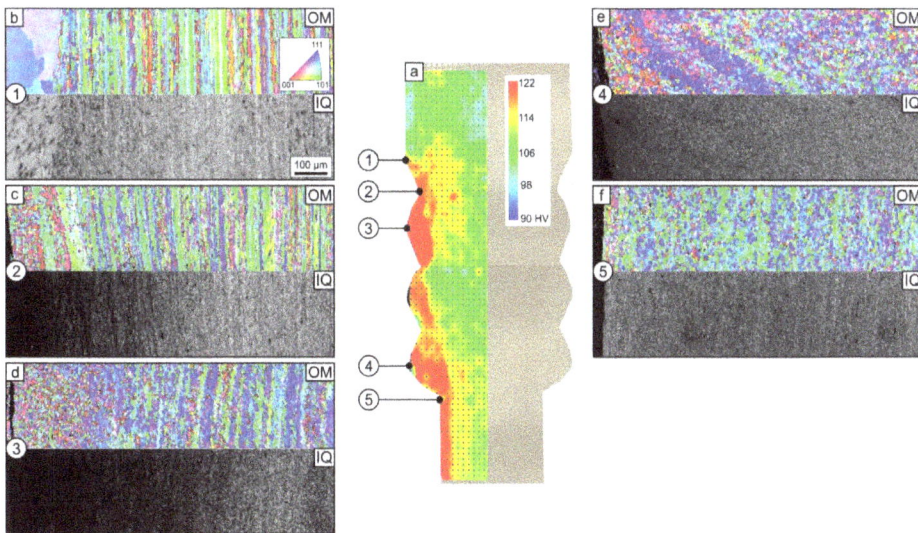

Figure 6. Hardness (**a**) and microstructure (**b–f**) of extruded material observed in a billet split lengthwise (step size 3 μm). Five distinct locations (1–5) for microstructural analysis, representing interesting stages of deformation history (see also Figure 3).

In Figure 7, a more detailed microstructural investigation of the fine-grained surface area is presented. SEM images and color-coded orientation maps from EBSD measurements of Points 2–5 are shown; the corresponding extrusion direction is highlighted in Figure 7a. Note that representative SEM images and the corresponding EBSD maps are taken from the same region but not from exactly the same area. With the help of the EBSD measurements, mean grain sizes and grain shape aspect ratios for the investigated points were derived (Table 3). Note that the white speckles in the SEM images are typical Fe-/Mn-rich precipitates [39] and are not further considered here. Figure 7a corresponds to the surface area at Point 2. As discussed earlier, a grain refinement due to SPD processing can be observed and the initially coarse-grained surface layer has been fully refined. The initial material at Point 1 exhibits an average grain size of about 150 μm in the surface area and 10.2 ± 2.3 μm in the interior region (see Table 3). The grain size at Point 2 is of ultrafine-grained (UFG) type with 0.79 ± 0.24 μm, and is homogeneously distributed. This result highlights that gradation extrusion is an effective process to provide the formation of numerous high-angle grain boundaries, resulting in a UFG microstructure. Note that the standard deviation of the grain size for all investigated locations is between 20 and 30% of the measured grain size, which is a characteristic value not originating from an uncertainty of the measurement but from the typical distribution of the grain size of SPD-processed materials. The corresponding grain shape (aspect ratio 2.7, see Table 3) remains almost unaffected compared with that in the initial structure (aspect ratio 3.0) since gradation extrusion introduces a shear deformation which also results in elongated grains.

Figure 7b shows the microstructure of the near-surface area at Point 3. The average grain size (0.60 ± 0.18) has been slightly reduced compared with the grain size at Point 2. Furthermore, the aspect ratio of the grains has also been slightly reduced to 2.5, indicating that the fraction of grains originating from the transformation of equiaxed dislocation cells into high-angle boundaries is increased.

The microstructure of the surface area at Point 4 (see Figure 7c) shows further changes: the former arrangement of elongated grains (see also Figure 6c,e) in the extrusion direction cannot be observed anymore. An ultrafine deformation microstructure with a grain size of 0.51 ± 0.16 μm can be observed with an aspect ratio of 2.1, which is slightly lower compared to those at Point 3. Interestingly, localized plastic deformation occurred at this stage. The corresponding SEM image reveals distinct shear localization and the presence of shear bands (see also Figure 8).

Figure 7d shows the microstructure of the near-surface area at Point 5, which represents the condition after the final reduction of the gradation extrusion. This final reduction leads again to an elongation of the grains parallel to the pressing direction. As a consequence of this extrusion step, the aspect ratio increases to 2.9. This is almost similar to the initial material, but compared to the initial material, the final grain size is significantly lower (0.66 ± 0.20 μm).

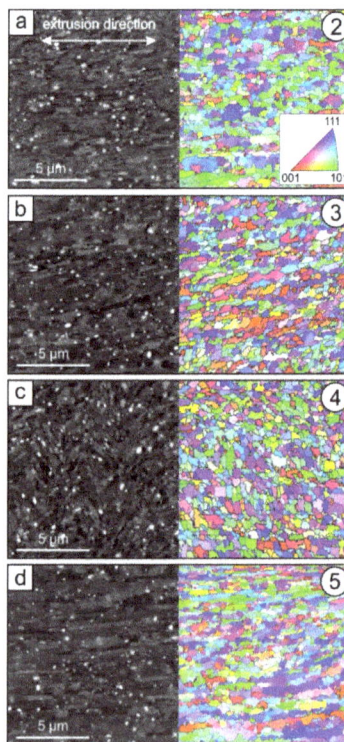

Figure 7. Microstructural details (**a–d**) of the near surface region at Points 2–5 (see also Figure 6) during gradation extrusion (step size 50 nm). Extrusion direction shown in (**a**) applies to all figures.

Table 3. Grain sizes and aspect ratios after different deformation steps, determined from EBSD data with a step size of 50 nm.

Location	Grain Size/μm	Aspect Ratio
1	10.20 ± 2.30	3.0
2	0.79 ± 0.24	2.7
3	0.60 ± 0.18	2.5
4	0.51 ± 0.16	2.1
5	0.66 ± 0.20	2.9

Figure 8 shows a SEM micrograph of a heterogeneously deformed region (see also Figure 7c). The alignment of elongated ultrafine grains (see dashed line in Figure 8) reveals the presence of two distinct band-like regions that have a width of several microns. An alternating arrangement of severely sheared regions next to regions with less strain is observed. These regions of localized straining are known to often act as initiation sites for cracking during further severe plastic deformation [40,41]. The absence of cracks in the billets after gradation extrusion indicates that microstructural gradients may also be beneficial in suppressing billet failure by shear localization.

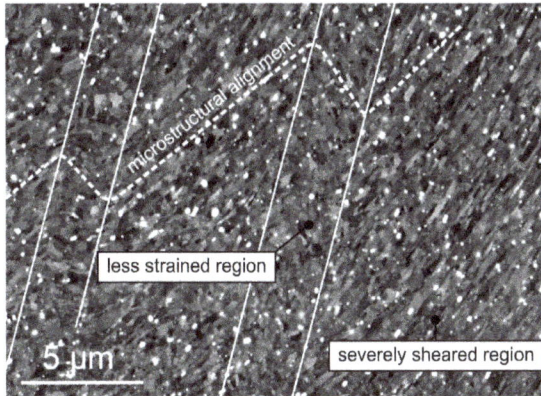

Figure 8. SEM micrograph of a heterogeneously deformed region at Point 4.

4. Discussion and Conclusions

With the help of an adapted die design that allows an interruption in processing and nondestructive billet removal, a systematic study of microstructural evolution during gradation extrusion was successfully performed. Our results highlight that hardness as well as grain size do not considerably change beyond a certain point (Point 3) in the deformation history. There are several microstructural fluctuations following the path of deformation in the near-surface area, most likely induced by a complex interplay of work hardening and grain refinement accompanied by temperature-driven effects like dynamic recovery and recrystallization. While grain size and grain shape undergo only slight changes beyond that certain point (Point 3), the arrangements of elongated grains and their preferred crystallographic orientation change towards a typical SPD microstructure, which has also been generated, for example, by multi-pass ECAP in other studies. These results show that gradation extrusion is a powerful tool to produce graded materials with partially ultrafine-grained microstructures. From the point of view of obtaining an ultrafine-grained surface layer with maximum hardness, only a small number of forming elements that induce significant amounts of shear deformation is needed. In this study, one concave and one convex forming element were demonstrated to be sufficient to produce a homogeneous ultrafine-grained surface area. However, from the point of view of achieving a full annihilation of the initial lamellar microstructure and texture, a larger number of forming elements is needed. A wider region of homogeneous microstructure is found after deformation by 3 concave and 3 convex forming elements with a subsequent reduction of the billet diameter. However, it is found that this large incremental deformation may result in locally heterogeneous microstructures near the billet surface caused by localization of deformation. Since localized deformation like shear banding can act as a mechanism of material failure, too many forming elements during gradation extrusion are found to be disadvantageous. Finally, considering economical aspects of processing, fewer forming elements are preferred since several processing parameters (e.g., pressing force and processing duration) are significantly reduced.

In closing, we comment on the relevance of the new process of gradation extrusion presented here compared to the already well-established SPD methods. ECAP provides comparatively large billets and large strains by multi-pass processing that typically results in homogeneous microstructures with sub-micron grain sizes. HPT is often of high scientific interest due to the possibility to introduce extremely large strains by multiple turns resulting in many cases in nanometer-range grains. However, a major disadvantage is the limited material volume which is processed in small cylindrical specimens. ARB is a suitable SPD method for sheet material and laminated structures often resulting in grain elongation and sub-micron grain sizes. Gradation extrusion, as presented in this study, is a powerful tool for tailor-made surface modifications of rod-shaped billets and enables, for example, the combination of a conventionally grained ductile center and an ultrafine-grained hard surface; as such, it is a complementary and valuable addition to the well-established SPD techniques, especially in terms of future practical applications.

Acknowledgments: The authors gratefully acknowledge funding by the German Research Foundation (Deutsche Forschungsgemeinschaft, DFG) through the Collaborative Research Center SFB 692 (projects C5, A1, A4 and Z2).

Author Contributions: Philipp Frint coordinated the experiments, performed mechanical testing, analyzed and discussed the results of the experiments, structured and wrote the manuscript. Markus Härtel analyzed and discussed the results of the experiments, structured and wrote the manuscript. René Selbmann performed the deformation experiments and FE simulation, supported writing the manuscript. Dagmar Dietrich performed the EBSD measurements, discussed the results and supported writing the manuscript. Markus Bergmann coordinated and performed the deformation experiments and FE simulation. Thomas Lampke supervised the EBSD measurements and discussed the results. Dirk Landgrebe supervised the deformation experiments and discussed the results. Martin F.-X. Wagner supervised the experiments and discussed the results, supported writing the manuscript.

Conflicts of Interest: The authors declare no conflict of interest.

References

1. Valiev, R.Z.; Islamgaliev, R.K.; Alexandrov, I.V. Bulk nanostructured materials from severe plastic deformation. *Prog. Mater. Sci.* **2000**, *45*, 103–189. [CrossRef]
2. Langdon, T.G. Twenty-five years of ultrafine-grained materials: Achieving exceptional properties through grain refinement. *Acta Mater.* **2013**, *61*, 7035–7059. [CrossRef]
3. Valiev, R.Z.; Langdon, T.G. Principles of equal-channel angular pressing as a processing tool for grain refinement. *Prog. Mater. Sci.* **2006**, *51*, 881–981. [CrossRef]
4. Segal, V.M.; Reznikov, A.E.; Drobyshevskiy, A.E.; Kopylov, V.I. Plastic working of metals by simple shear. *Russ. Metall.* **1981**, *1*, 99–105.
5. Frint, S.; Hockauf, M.; Frint, P.; Wagner, M.F.X. Scaling up segal's principle of equal-channel angular pressing. *Mater. Des.* **2016**, *97*, 502–511. [CrossRef]
6. Frint, P.; Hockauf, M.; Halle, T.; Strehl, G.; Lampke, T.; Wagner, M.F.X. Microstructural features and mechanical properties after industrial scale ECAP of an Al-6060 alloy. *Mater. Sci. Forum* **2011**, *667–669*, 1153–1158. [CrossRef]
7. Saito, Y.; Tsuji, N.; Utsunomiya, H.; Sakai, T.; Hong, R.G. Ultra-fine grained bulk aluminum produced by accumulative roll-bonding (ARB) process. *Scr. Mater.* **1998**, *39*, 1221–1227. [CrossRef]
8. Smirnova, N.A.; Levit, V.I.; Pilyugin, V.P.; Kuznetsov, R.I.; Davydova, L.S.; Sazonova, V.A. Evolution of structure of fcc single crystals during strong plastic deformation. *Phys. Metals Metall.* **1986**, *61*, 127–134.

9. Zhilyaev, A.P.; Langdon, T.G. Using high-pressure torsion for metal processing: Fundamentals and applications. *Prog. Mater. Sci.* **2008**, *53*, 893–979. [CrossRef]
10. Frint, P.; Hockauf, M.; Dietrich, D.; Halle, T.; Wagner, M.F.X.; Lampke, T. Influence of strain gradients on the grain refinement during industrial scale ECAP. *Mater. Werkst.* **2011**, *42*, 680–685. [CrossRef]
11. Frint, P.; Hockauf, M.; Halle, T.; Wagner, M.F.X.; Lampke, T. The role of backpressure during large scale equal-channel angular pressing. *Mater. Werkst.* **2012**, *43*, 668–672. [CrossRef]
12. Xu, C.; Langdon, T.G. Influence of a round corner die on flow homogeneity in ECA pressing. *Scr. Mater.* **2003**, *48*, 1–4. [CrossRef]
13. Yoon, S.C.; Quang, P.; Hong, S.I.; Kim, H.S. Die design for homogeneous plastic deformation during equal channel angular pressing. *J. Mater. Process. Technol.* **2007**, *187–188*, 46–50. [CrossRef]
14. Barber, R.E.; Dudo, T.; Yasskin, P.B.; Hartwig, K.T. Product yield for ecae processing. *Scr. Mater.* **2004**, *51*, 373–377. [CrossRef]
15. Iwahashi, Y.; Furukawa, M.; Horita, Z.; Nemoto, M.; Langdon, T.G. Microstructural characteristics of ultrafine-grained aluminum produced using equal-channel angular pressing. *Metall. Mater. Trans. A* **1998**, *29*, 2245–2252. [CrossRef]
16. Vorhauer, A.; Pippan, R. On the homogeneity of deformation by high pressure torsion. *Scr. Mater.* **2004**, *51*, 921–925. [CrossRef]
17. Xu, C.; Horita, Z.; Langdon, T.G. The evolution of homogeneity in processing by high-pressure torsion. *Acta Mater.* **2007**, *55*, 203–212. [CrossRef]
18. Estrin, Y.; Vinogradov, A. Extreme grain refinement by severe plastic deformation: A wealth of challenging science. *Acta Mater.* **2013**, *61*, 782–817. [CrossRef]
19. Azushima, A.; Kopp, R.; Korhonen, A.; Yang, D.Y.; Micari, F.; Lahoti, G.D.; Groche, P.; Yanagimoto, J.; Tsuji, N.; Rosochowski, A.; et al. Severe plastic deformation (SPD) processes for metals. *CIRP Ann. Manuf. Technol.* **2008**, *57*, 716–735. [CrossRef]
20. Dobatkin, S.V.; Szpunar, J.A.; Zhilyaev, A.P.; Cho, J.Y.; Kuznetsov, A.A. Effect of the route and strain of equal-channel angular pressing on structure and properties of oxygen-free copper. *Mater. Sci. Eng. A* **2007**, *462*, 132–138. [CrossRef]
21. Hoseini, M.; Meratian, M.; Toroghinejad, M.R.; Szpunar, J.A. Texture contribution in grain refinement effectiveness of different routes during ECAP. *Mater. Sci. Eng. A* **2008**, *497*, 87–92. [CrossRef]
22. Gholinia, A.; Prangnell, P.B.; Markushev, M.V. The effect of strain path on the development of deformation structures in severely deformed aluminium alloys processed by ECAE. *Acta Mater.* **2000**, *48*, 1115–1130. [CrossRef]
23. Niehuesbernd, J.; Müller, C.; Pantleon, W.; Bruder, E. Quantification of local and global elastic anisotropy in ultrafine grained gradient microstructures, produced by linear flow splitting. *Mater. Sci. Eng. A* **2013**, *560*, 273–277. [CrossRef]
24. Neugebauer, R.; Sterzing, A.; Bergmann, M. Mechanical properties of the AlSi1MgMn aluminium alloy (AA6082) processed by gradation rolling. *Mater. Werkst.* **2011**, *42*, 593–598. [CrossRef]
25. Richert, M.; Petryk, H.; Stupkiewicz, S. Grain refinement in AlMgSi alloy during cyclic extrusion—Compression: Experiment and modelling *Arch. Metall. Mater.* **2007**, *52*, 49 54.
26. Lampke, T.; Dietrich, D.; Nickel, D.; Bergmann, M.; Zachäus, R.; Neugebauer, R. Controlled grain size distribution and refinement of an EN AW-6082 aluminium alloy. *Int. J. Mat. Res.* **2011**, *102*, 1–5. [CrossRef]
27. Neugebauer, R.; Sterzing, A.; Selbmann, R.; Zachäus, R.; Bergmann, M. Gradation extrusion—Severe plastic deformation with defined gradient. *Mater. Werkst.* **2012**, *43*, 582–588. [CrossRef]
28. Landgrebe, D.; Sterzing, A.; Schubert, N.; Bergmann, M. Influence of die geometry on performance in gradation extrusion using numerical simulation and analytical calculation. *CIRP Ann. Manuf. Technol.* **2016**, *65*, 269–272. [CrossRef]
29. Zachäus, R.; Bergmann, M. Verfahren und Vorrichtung Zur Korngrössenbeeinflussung Eines Werkstückes Sowie Werkstück. DE 102012006952 B4, 2012.
30. Neugebauer, R.; Bergmann, M. Local severe plastic deformation by modified impact extrusion process. In *Steel Research International*; WILEY-VCH Verlag GmbH & Co.: Weinheim, Germany, 2012.
31. Frint, P.; Wagner, M.F.X.; Weber, S.; Seipp, S.; Frint, S.; Lampke, T. An experimental study on optimum lubrication for large-scale severe plastic deformation of aluminum-based alloys. *J. Mater. Process. Technol.* **2017**, *239*, 222–229. [CrossRef]

32. Baumgarten, J.; Bunk, W.; Luecke, K. Extrusion textures in Al-Mg-Si alloys—1. Round extrusions. *Z. Metall. Mater. Res. Adv. Tech.* **1981**, *72*, 75–81.

33. Sheppard, T.; Parson, N.C.; Zaidi, M.A. Dynamic recrystallization in Al-7Mg alloy. *Metal Sci.* **1983**, *17*, 481–490. [CrossRef]

34. Sheppard, T. *Extrusion of Aluminium Alloys*, 1st ed.; Springer: Berlin, Germany, 1999.

35. Dietrich, D.; Berek, H.; Schulze, A.; Scharf, I.; Lampke, T. EBSD and STEM on aluminum alloys subjected to severe plastic deformation. *Prakt. Metall. Prac. Metall.* **2011**, *48*, 136–150. [CrossRef]

36. Berndt, N.; Frint, P.; Böhme, M.; Wagner, M.F.X. Microstructure and mechanical properties of an AA6060 aluminum alloy after cold and warm extrusion. *Mater. Sci. Eng. A* **2017**, *707*, 717–724. [CrossRef]

37. Schwartz, A.J.; Kumar, M.; Adams, B.L.; Field, D.P. *Electron Backscatter Diffraction in Materials Science*; Springer: Berlin, Germany, 2009.

38. Yin, Q.; Zillmann, B.; Suttner, S.; Gerstein, G.; Biasutti, M.; Tekkaya, A.E.; Wagner, M.F.-X.; Merklein, M.; Schaper, M.; Halle, T.; Brosius, A. An experimental and numerical investigation of different shear test configurations for sheet metal characterization. *Int. J. Solids Struct.* **2014**, *51*, 1066–1074. [CrossRef]

39. Birol, Y. The effect of processing and Mn content on the T5 and T6 properties of AA6082 profiles. *J. Mater. Process. Technol.* **2006**, *173*, 84–91. [CrossRef]

40. Antolovich, S.D.; Armstrong, R.W. Plastic strain localization in metals: Origins and consequences. *Prog. Mater. Sci.* **2014**, *59*, 1–160. [CrossRef]

41. Lee, W.B.; Chan, K.C. A criterion for the prediction of shear band angles in F.C.C. Metals. *Acta Metall. Mater.* **1991**, *39*, 411–417. [CrossRef]

metals

MDPI

Article

Effect of Nitric and Oxalic Acid Addition on Hard Anodizing of AlCu₄Mg₁ in Sulphuric Acid

Maximilian Sieber [1], Roy Morgenstern [2,*], Ingolf Scharf [2] and Thomas Lampke [2]

[1]   Lockwitzgrund 123a, D-01731 Kreischa, Germany; maximiliansieber@yandex.com
[2]   Materials and Surface Engineering Group, Chemnitz University of Technology, D-09107 Chemnitz, Germany; ingolf.scharf@mb.tu-chemnitz.de (I.S.); thomas.lampke@mb.tu-chemnitz.de (T.L.)
*   Correspondence: roy.morgenstern@mb.tu-chemnitz.de; Tel.: +49-371-531-32818

Received: 15 December 2017; Accepted: 13 February 2018; Published: 17 February 2018

Abstract: The anodic oxidation process is an established means for the improvement of the wear and corrosion resistance of high-strength aluminum alloys. For high-strength aluminum-copper alloys of the 2000 series, both the current efficiency of the anodic oxidation process and the hardness of the oxide coatings are significantly reduced in comparison to unalloyed substrates. With regard to this challenge, recent investigations have indicated a beneficial effect of nitric acid addition to the commonly used sulphuric acid electrolytes both in terms of coating properties and process efficiency. The present work investigates the anodic oxidation of the AlCu₄Mg₁ alloy in a sulphuric acid electrolyte with additions of nitric acid as well as oxalic acid as a reference in a full-factorial design of experiments (DOE). The effect of the electrolyte composition on process efficiency, coating thickness and hardness is established by using response functions. A mechanism for the participation of the nitric acid additive during the oxide formation is proposed. The statistical significance of the results is assessed by an analysis of variance (ANOVA). Eventually, scratch testing is applied in order to evaluate the failure mechanisms and the abrasion resistance of the obtained conversion coatings.

Keywords: anodic oxidation; additives; aluminum alloy AlCu₄Mg₁; nitric acid; oxalic acid; hardness; porosity; energy efficiency; scratch resistance

---

1. Introduction

The anodic oxidation process is a suitable means for the surface refinement of aluminum and its alloys. The formation of an oxide ceramic coating under anodic polarization in an acidic electrolyte leads to an increased corrosion and wear resistance, enhances haptic-visual properties and depending on the process regime, provides certain other surface property alterations like electrical insulation. Anodic oxide coatings with a particularly low porosity and therefore high hardness and abrasion resistance can be achieved by anodizing in sulfuric acid electrolytes at low temperatures beneath 5 °C due to the reduced chemical dissolution of the oxide. However, the so-called "hard anodizing" process itself is costly and demands the input of substantial amounts of electrical energy for both the process itself and the temperature control of the electrolyte. Another means of reducing the coating porosity lies in the addition of organics to the commonly used sulphuric acid electrolyte. Giovanardi et al. [1] proved that organic additions, e.g., glycolic acid, oxalic acid and glycerol, limit the chemical dissolution of the pore walls by adsorbing at the oxide-solution interface. Although the current efficiency of the process as well as the coatings' hardness and wear resistance are improved [2], the overall energy consumption is often in the same way increasing, since the mentioned additives increase the process voltage and thus thwart the alleged efficiency improvement [3].

A second challenge for the production of functional oxide coatings arises from the substrate influence. Being conversion coatings, the properties of the alumina coatings produced by anodizing are inevitably affected by the substrate alloy. While improving the material strength, alloying elements like copper have a detrimental effect on the aluminum oxide formation. From the thermodynamic view, the oxidation of aluminum atoms is significantly preferred to the oxidation of fine dispersed copper atoms due to the more negative Gibbs free energy per equivalent for the formation of aluminum oxide [4]. This leads to copper enrichment at the substrate coating interface [4]. Hashimoto et al. describe the formation of the θ′-phase (Al_2Cu) within the copper-enriched layer [5]. These nanoscale copper-rich phases are oxidized at technologically relevant anodic potentials of more than 4 V [6]. Because of the semiconductive properties of copper oxide, this process is accompanied by oxygen evolution [4–6]. As electrical charge is consumed during this side reaction, the current efficiency of oxide growth is significantly reduced. Moreover, additional voids can be observed along the pore channels as the enrichment, oxidation and oxygen evolution process repeats in regular time intervals. Apart from this, the oxidation of intermetallic phases in the substrate alloy leads to micron scale defects in the conversion coating. Ma et al. [7] differentiate between copper-rich phases which are preferentially dissolved leaving voids and iron-rich phases which hinder the conversion process leaving highly porous volumes and voids in the coating. Because of the increased porosity, anodic oxide coatings on aluminum-copper alloys exhibit lower hardness and abrasion resistance.

Recent investigations indicate a beneficial effect of the addition of nitric acid to a sulphuric acid electrolyte on the performance of the conversion coatings produced on the popular alloy $AlCu_4Mg_1$ (EN-AW 2024) [3]. In the same time, the process voltage of the hard-anodizing process is decreased, which leads to a lower energy consumption for the oxide production. The current study focuses on two major aspects: (1) determination of the effect of the nitric acid addition on the characteristics of both the hard-anodizing process and the produced coatings in dependence of the nitric acid concentration in the sulphuric acid electrolyte; (2) investigation into the mechanism behind the effects of nitric acid. Therefore, alongside the determination of the thickness, hardness and scratch-resistance of the produced coatings, as well as the quantification of the current efficiency and the energy consumption of the hard-anodizing process, the microstructure and composition of the produced coatings is investigated. Thus, it shall be clarified whether or not nitric acid is a suitable additive to improve the anodic oxidation especially of copper-alloyed aluminum substrates.

2. Materials and Methods

2.1. Anodizing Process

The alloy EN AW-2024 T3 (nominal composition is given in Table 1) served as substrate material for the anodic oxidation. It was supplied as sheet metal (Q-Lab, Westlake, OH, USA) and was used with dimensions of 50 mm × 25 mm × 1.5 mm. The samples were etched in 3 wt % sodium hydroxide at 50 °C for 5 min and pickled in 1:1 nitric acid at room temperature for 30 s. After each step, the samples were rinsed under deionized water. The anodic oxidation was carried out in 20 vol % sulphuric acid (corresponds to approx. 3.75 mol/L, Merck, Darmstadt, Germany) with additions of 0.4 mol/L and 0.8 mol/L nitric acid and 0.2 mol/L oxalic acid (as oxalic acid dihydrate, Merck). The electrolyte, which had a volume of 2 L, was maintained at a temperature of 5 ± 2 °C (being typical of hard-anodizing) throughout the process with a thermostat. The electrolyte was constantly stirred with a rod agitator (300 rpm). A pe1028 power station (Plating Electronic, Sexau, Germany) served as the power source. Current and voltage signals were logged internally with a sampling rate of one sample per second. The anodizing process was carried out in galvanostatic mode with a current density of 3 A/dm². The process was terminated after 45 min.

Table 1. Chemical composition of alloy EN AW-2024.

Element	Al	Cu	Mg	Mn	Si	Fe	Cr	Zn	Ti
wt %	balance	3.9–4.9	1.2–1.8	0.3–0.9	≤0.5	≤0.5	≤0.1	≤0.25	≤0.15

2.2. Coating and Process Characterization

The effects of the nitric and oxalic acid addition were considered with regard to several process and coating properties. The electrical energy consumption during the anodizing process W_{el} was calculated by integrating the product of current density and voltage over the process time. The thickness s of the produced coatings was obtained by eddy current measurement (Fischerscope MMS, Fischer, Sindelfingen, Germany). The values were validated on cross sections of the coatings. The mass of the anodized samples was determined before and after dissolution of the alumina in chromic/phosphoric acid (35 mL/L phosphoric acid + 20 g/L chromium(VI)oxide) at 60 °C for 4 h using a X1003S balance (Mettler Toledo, Gießen, Germany). The said solution dissolves alumina and does not attack aluminum. Preliminary tests also showed no attack on the used AlCu$_4$Mg$_1$ alloy. The specific mass m was obtained by dividing the coating mass (which is equal to the mass loss after dissolution) by the surface area. According to Faraday´s law, the specific mass of alumina produced in a galvanostatic process carried out at 3 A/dm^2 for 45 min amounts to 1426.6 mg/dm^2 theoretically. The current efficiency η was calculated by dividing the mass loss obtained from the exposition in chromic/phosphoric acid by the theoretical coating mass. In the same way, the energy efficiency ε was calculated by dividing the electrical energy consumption during the process by the actual oxide mass, which gives a measure of how much energy is required for the formation of a certain coating mass ([ε] = J/mg). Under the assumption that the theoretical alumina density ρ (3.95 g/cm^3) is valid for the anodic alumina produced under the described conditions, the porosity p of the coatings was calculated by applying the following formula using the specific coating mass m and the coating thickness s:

$$p = 1 - \frac{m}{s \cdot \rho} \tag{1}$$

The hardness of the coatings was obtained from instrumented indentation tests at different locations of the coatings' cross sections with a Berkovich indenter (UNAT, Asmec, Dresden, Germany). A load of 5 mN was applied (load time 10 s, hold time 5 s, unload time 4 s). The distance between each indent and the substrate/coating interface was registered. The resulting hardness profiles were approximated by an exponential function, which represents the decline of the hardness H with increasing distance d from the substrate.

$$H = H_0 \cdot (H^*)^d \tag{2}$$

The lowest deviation between the experimental results and Equation (2) were obtained for a parameter $H_0 = 4400$ N/mm^2 over the complete set of data. The value H^* represents the hardness decline. For a value $H^* = 1$, there is no hardness decline at all. With decreasing H^*, the hardness decline gets more pronounced. With the distance d in microns, H^* was in a range of approx. 0.96 to 0.99 for the present set of samples. The hardness of the coatings in a distance of 20 μm to the substrate was considered as a reference value in the DOE exploitation.

2.3. DOE-Set Up and Exploitation

A full factorial design was used to assess the effects of the additives on the coating and process characteristics. The steps for the nitric acid concentration were 0 mol/L, 0.4 mol/L and 0.8 mol/L, while the oxalic acid concentration was 0 or 0.2 mol/L. The full factorial design thus included 12 different electrolytes. For each of the electrolytes, three samples were produced independently and all the properties were determined for these samples, with the exception of the hardness, which was measured on the cross section of only two samples for each of the electrolytes with approx. 20 indents across the coating cross section. The values obtained for each of the process and coating properties were used as input values for a model, which was quadratic with regard to the nitric acid concentration and linear with regard to the oxalic acid concentration. A preliminary consideration of the statistical significance of the parameter effects on the properties showed no statistical significance for the interaction of the oxalic and nitric acid additives. Interaction terms have therefore not been considered and the response function for a generic property g (e.g., coating thickness s, energy efficiency ε, hardness H_{20}) with the coefficients a_i (i = 1, 2, 3, 4) and the concentrations of oxalic and nitric acid was chosen as follows:

$$g = a_1 + a_2 \cdot c_{oxalic} + a_3 \cdot c_{nitric}^2 + a_4 \cdot c_{nitric} \qquad (3)$$

The coefficients were determined by least-square fitting of the model to the obtained results for each parameter. Thus, a quantitative relation of each property and the additive concentration was obtained. The quality of the model was checked for each of the parameters by comparing the model predictions g_{pred} with the measured properties g_{meas} ($\overline{g}_{meas}$ being the mean measured value) via the following function:

$$R^2 = 1 - \frac{\sum \left(g_{meas} - g_{pred} \right)^2}{\sum (g_{meas} - \overline{g}_{meas})^2} \qquad (4)$$

2.4. Microstructure Characterisation

Metallographic cross sections were prepared by grinding on SiC paper and polishing with diamond suspension accomplished by a finish using a silicon oxide polishing suspension. Before the scanning electron microscopy (SEM) investigations, the cross-sections were cleaned thoroughly, dried at 60 °C for at least 4 h and carbon coated in order to avoid sample charging. The microstructure was investigated by SEM (LEO 1455VP, Zeiss, Jena, Germany). Both secondary electron (SE, topography contrast) and backscattered electron (BSD, element contrast) detectors were applied. For the quantitative analysis of the submicron and micron scale porosity, BSD pictures were assembled in order to get a representative impression of the coating microstructure over a length of more than 300 microns. The porosity was calculated via the grey-scale of pixels, whereby pixels with a grey-scale lower than a suitable threshold value were counted and the sum was put in relation with the total number of pixels.

2.5. Scratch Testing and Profilometry

Scratch tests were performed with a Revetest-RST scratch tester (CSM Instruments, Peseux, Switzerland) using a Rockwell diamond cone indenter (radius 200 µm) in order to evaluate the coatings' adhesion and the two-body abrasion resistance of the coatings. The sample was moved relative to the indenter with a speed of 2.5 mm/min. A prescan and a postscan were conducted at a small normal load of 0.9 N in order to calculate the remaining scratch depth. During scratch testing, the tangential force and the acoustic emission were recorded. For the quantification of the adhesive failure, the normal load was linearly increased within a scratch length of 10 mm from 1 to 100 N. This procedure will be referred to as "progressive scratch testing". The first occurrence of adhesive failure was determined by the help of the remaining scratch depth and by optical examination of the scratch. The force, at which the coating failed, meaning the breakthrough to the substrate, will be referred to as the critical force Fc. For the quantification of the abrasion resistance, scratch tests with a length of 5 mm were conducted at a constant normal load of 10 N. This procedure will be referred to as "constant scratch testing". The cross-section profile of each constant scratch was recorded at three positions using tactile measurement (T8000, Jenoptik, Jena, Germany). The software Turbo Wave was applied for the levelling of the profile and the calculation of the cross-section area. The scratch energy density W_R was calculated using the mean value of the tangential force F_t the mean value of the cross section area A and the scratch length l according to Equation (5).

$$W_R = \frac{F_t \cdot l}{A \cdot l} \qquad (5)$$

Two progressive scratch tests and three constant scratch tests were performed on each of two anodized samples per anodizing condition.

3. Results

3.1. Energy Efficiency

The voltage transients for the anodic oxidation process in the sulphuric acid electrolyte with different amounts of the oxalic and nitric acid additives differ significantly. In Figure 1, four representative voltage transients and the associated standard deviations are shown for the galvanostatic process (3 A/dm^2). In principal, each of the curves shows the typical voltage evolution during a galvanostatic anodizing process. The process initiation includes a steep voltage increase within the first seconds, which is attributed to the barrier layer formation and a subsequent decline of the voltage, and marks the beginning of the pore formation. Afterwards, the voltage grows at a slower rate throughout the process, which is attributed to the thickening of the porous part of the oxide coating. As can be seen in the diagram, the voltage amounts to about 23 V for the base electrolyte without additives after the process initiation. The oxalic acid additive does not affect the voltage level after the process initiation to a technologically relevant amount. However, the nitric acid additive leads to a significantly lower voltage in the first minutes of the process. With increasing process time, the voltage remains almost constant for the electrolyte without additives, while both additives lead to a significant growth of the voltage. For oxalic acid, the slope of the voltage curve is especially steep at a process time of around 25 min, while the nitric acid additive leads to a more or less constant voltage growth throughout the process. A higher amount of nitric acid addition leads to a further decrease of the voltage level after the process initiation, while the slope of the voltage in the further course of the process gets more pronounced. Therefore, the overall electrical energy consumption during the process is, of course, affected. In the sulphuric acid electrolyte without additives, the electrical energy turnover of the anodic oxidation process itself amounts to approx. 54.5 ± 0.7 Wh/dm^2. The addition of oxalic acid leads to an increase of W_{el} to 73 ± 6 Wh/dm^2. In contrast, the addition of 0.4 mol/L and 0.8 mol/L nitric acid decrease the electrical energy consumption slightly to values of 52.2 ± 1.1 Wh/dm^2 and 51.6 ± 0.6 Wh/dm^2, respectively. All the values are summarized in Table 2.

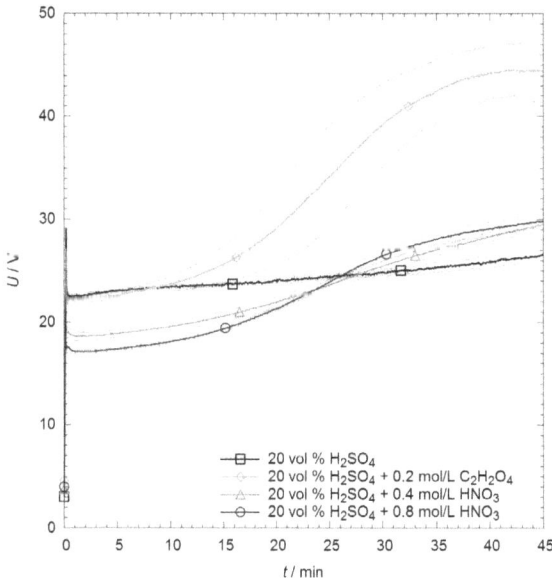

Figure 1. Voltage curves for the galvanostatic anodic oxidation process (current density: 3 A/dm^2) in the sulphuric acid base electrolyte with and without various additions. The thin lines represent the 66% confidence intervals, i.e., the standard deviation.

Table 2. Results for the considered process and coating properties for all the combinations of additions of oxalic and nitric acid to the sulphuric acid base electrolyte.

Property	Symbol	Unit	Oxalic/Nitric Acid Addition in mol/L					
			0/0	0/0.4	0/0.8	0.2/0	0.2/0.4	0.2/0.8
Coating thickness	s	μm	42 ± 5	44 ± 2	48 ± 2	42 ± 3	45 ± 3	48 ± 3
Specific mass	m	mg/dm^2	855 ± 16	964 ± 5	1025 ± 4	850 ± 10	971 ± 9	1026 ± 14
El. energy Consumption	W_{el}	Wh/dm^2	54.5 ± 0.6	52.2 ± 0.9	51.6 ± 0.5	73 ± 5	65.6 ± 1.0	67 ± 5
Current eff.	η	%	59.9 ± 1.2	67.6 ± 0.4	71.9 ± 2.7	59.6 ± 0.7	68.0 ± 0.6	72.0 ± 1.0
Energy eff.	ε	J/mg	230 ± 7	195 ± 3	181 ± 2	311 ± 22	243 ± 6	234 ± 14

The fit of the quadratic response function to the results leads to the coefficients shown in Table 3. Consequently, the graph represented in Figure 2a depicts the influence of the additives on the electrical energy consumption. The deviation of the predicted values obtained from the response function and the measured results is shown in Figure 3a. It is clearly visible that the oxalic acid addition has the biggest influence on the electrical energy consumption. The slight decrease of W_{el} by nitric acid occurs independently of the oxalic acid addition. The response function represents the measured values well, except for the highest values obtained at an oxalic and nitric acid concentration of 0.2 mol/L and 0 mol/L, respectively, which are underestimated by the model. As can be seen in Figure 2, the voltage curve for this electrolyte shows a comparatively big standard deviation, which directly propagates into the values for the electrical energy consumption. To evaluate the efficiency of the anodic oxidation, the coating mass is considered. Referring to the results shown in Table 2 and to the graphical representation in Figure 2b, it becomes clear that the addition of oxalic acid has no effect on the mass of the produced oxide coatings. Meanwhile, the addition of nitric acid leads to an increase of the produced oxide mass. The experimental results are well represented by the response function (Figure 3b).

Table 3. Coefficients of the response function for representation of the properties in dependence of the oxalic and nitric acid addition to the sulphuric acid base electrolyte.

Property	Symbol	Unit	a_1	a_2	a_3	a_4	R^2
Coating thickness	s	μm	40	3	2.3	6	0.96
Specific mass	m	mg/dm^2	900	-0.7	-180	400	0.98
El. energy consumption	W_{el}	Wh/dm^2	50	80	9	-11	0.87
Current efficiency	η	%	0.6	0.006	-0.12	0.25	0.98
Energy efficiency	ε	J/mg	230	270	70	-120	0.86

On the basis of this finding, two different routes shall be further pursued. The first route addresses the anodic oxidation process, namely its current and energy efficiency. The second route addresses the coating properties, namely the thickness, porosity, hardness and coating adhesion. With regard to the process, the obtained values of the oxide mass allow the calculation of the current efficiency η, i.e., how much of the overall charge turnover actually contributes to oxide formation, and the energy efficiency ε, i.e., how much energy is used for the formation of a certain amount of oxide. Since all the samples were produced in galvanostatic mode with a process time of 45 min and therefore with a constant charge turnover, the increase of the oxide mass by the addition of nitric acid into the electrolyte is directly reflected by an increased current efficiency (Table 2), while the oxalic acid additive does not affect neither of them (Figure 2c). The correlation between the values predicted by the response function and the experimental results is high (Figure 3c). In contrast, the energy efficiency ε increases significantly by the addition of oxalic acid, since the electrical energy consumption is increased without any change of the oxide mass (Figure 2d). That means, that more energy is needed for the production of a certain amount of oxide. The addition of nitric acid, meanwhile, decreases the energy efficiency, because the oxide mass is increased and the electrical energy consumption decreases at the same time. Consequently, less energy is needed for the production of a certain amount of oxide. The correlation between the predicted energy effiency and the measured values is compromised by the same error as the electrical energy consumption, so that especially the values for the electrolyte comprising only the oxalic acid additive are underestimated (Figure 3d). The coating thickness is hardly affected by the addition of oxalic acid, while it increases by approx. 10% after the addition of 0.8 mol/L nitric acid to the sulphuric acid electrolyte (Figure 2e). In the considered parameter range, the response function allows the prediction of the coating thickness with high accuracy (Figure 3e). Related to the results obtained in the base electrolyte without any additives, the increase of the coating thickness with increasing nitric acid concentration is stronger than the increase of the oxide mass. Under the assumption of a constant density of the amorphous alumina, this indicates an increasing coating porosity.

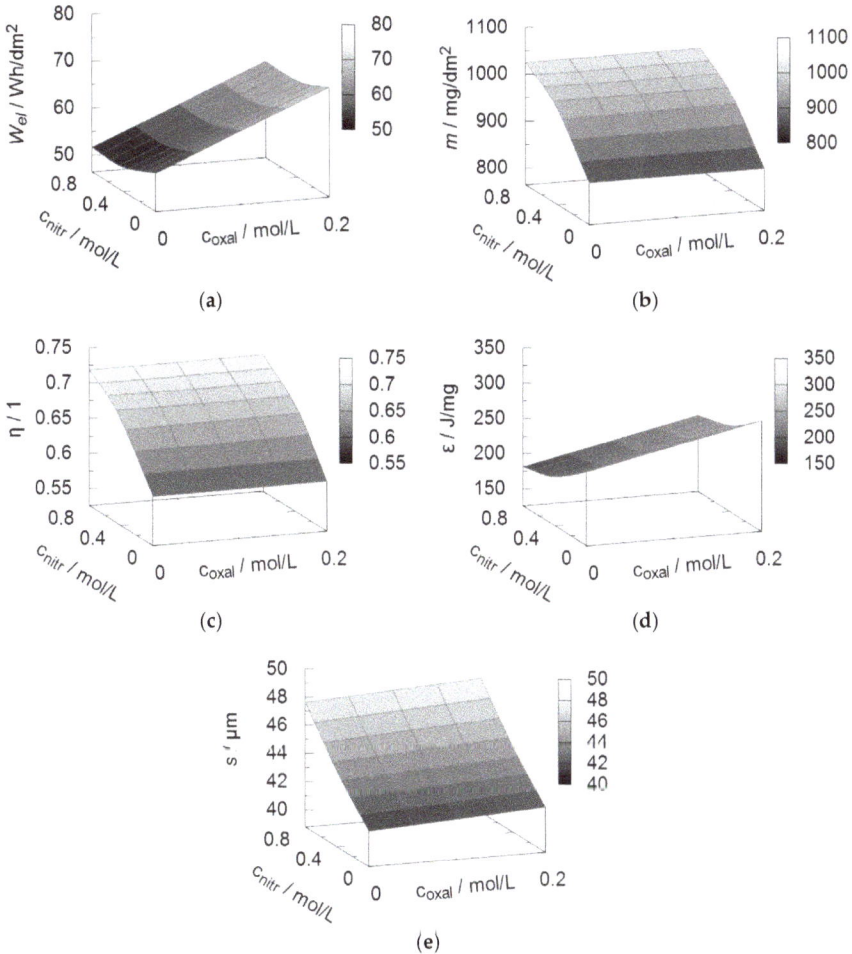

Figure 2. Effect of the addition of oxalic and nitric acid on the coating and process properties as predicted by the response function, (a) electrical energy consumption W_{el}, (b) oxide mass m, (c) current efficiency η, (d) energy efficiency ε, (e) coating thickness s.

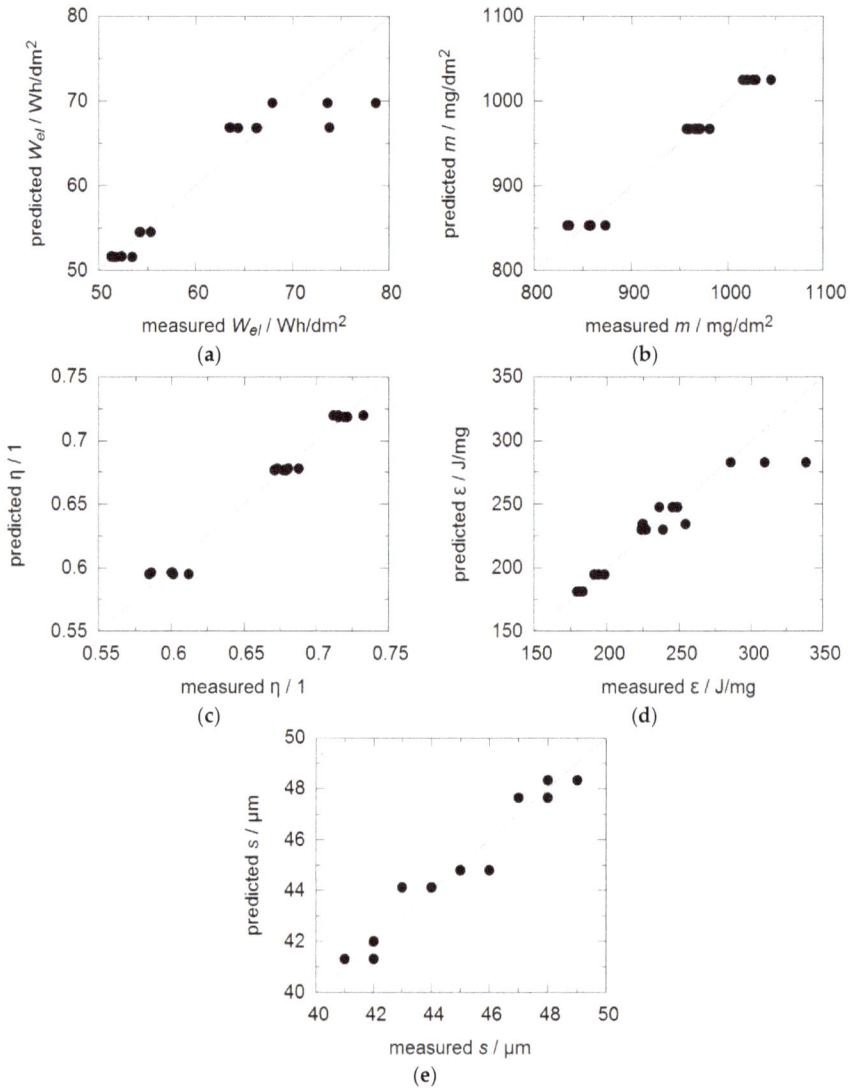

Figure 3. Comparison between values predicted by the response function and measured values for (**a**) electrical energy consumption W_{el}, (**b**) oxide mass m, (**c**) current eciency η, (**d**) energy efficiency ε, (**e**) coating thickness s.

3.2. Coating Porosity and Hardness

The overall coating porosity shows a minimum at a nitric acid concentration of approx. 0.4 mol/L according to the response function as represented in Figure 4a and Table 4. The coefficients of the quadratic response function are summarized in Table 5. At a constant nitric acid concentration, the porosity always slightly increases with the addition of oxalic acid. For the porosity, low values (45% and smaller) tend to be overestimated by the model while the higher values (around 50%) are underestimated (Figure 5a). As an additional measure for the compactness of the coatings, the hardness is considered. Generally, the size of the hardness indents is in the micrometer range and is thus in a greater order of magnitude compared to the pore channels and the periodically occurring voids. As can be seen from Figure 4c and Table 4, an increasing nitric acid concentration leads to a stronger increase of the hardness H_{20} and the hardness decline H^* in comparison with the oxalic acid addition. The maximum hardness H_{20} can be observed for the combined addition of 0.4 mol/L nitric acid and 0.2 mol/L oxalic acid. The further increase of the nitric acid concentration to 0.8 mol/L at an oxalic acid concentration of 0.2 mol/L leads to a reduced hardness H_{20}. A similar behavior can be observed for the hardness decline H^*, however, the decrease of the hardness for the highest nitric acid concentration is pronounced more significantly by the hardness decline. For the nitric acid concentration, the found effect on the hardness is in accordance with the estimation of the porosity, which showed a decreasing porosity for the addition of 0.4 mol/L nitric acid and again an increase for the addition of 0.8 mol/L nitric acid. However, at a constant nitric acid concentration, the porosity always increases by the addition of oxalic acid. For 0.0 mol/L and 0.4 mol/L nitric acid, this means that both porosity and hardness are increasing simultaneously. This phenomenon will be discussed later under the consideration of the coating microstructure.

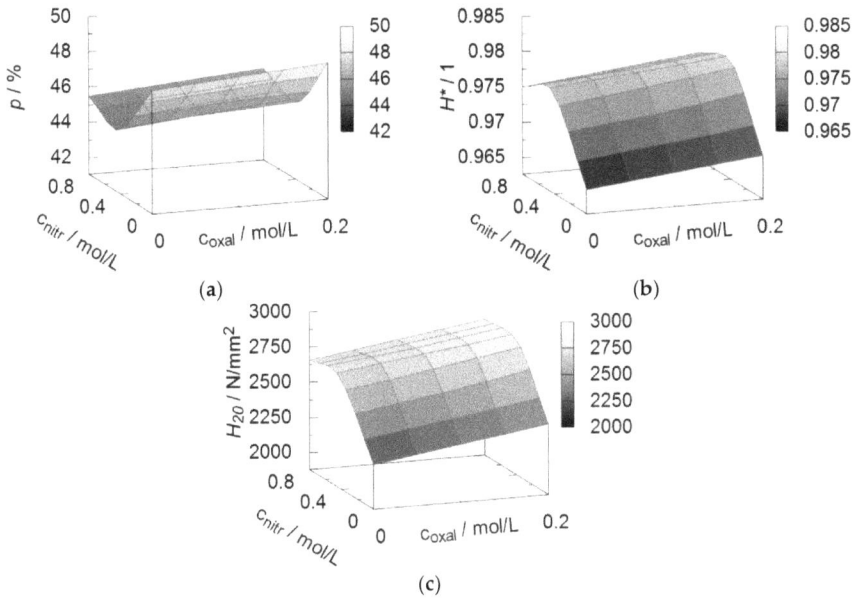

Figure 4. Effect of the addition of oxalic and nitric acid on the coating and process properties as predicted by the response function, (**a**) porosity p, (**b**) hardness decline H^*, (**c**) hardness in a distance of 20 microns from the substrate/oxide interface.

Table 4. Results for porosity-related coating properties for all the combinations of additions of oxalic and nitric acid to the sulphuric acid base electrolyte.

Property	Symbol	Unit	Oxalic/Nitric Acid Addition in mol/L					
			0/0	0/0.4	0/0.8	0.2/0	0.2/0.4	0.2/0.8
Porosity	p	%	48.0 ± 1.2	44.1 ± 0.8	45.6 ± 0.7	48.8 ± 0.6	45.8 ± 0.3	46.2 ± 0.2
Hardness	H_{20}	N/mm^2	2200	2600	2700	2300	3000	2800
Hardness decline	H^*		0.966	0.975	0.975	0.969	0.981	0.978

Table 5. Coefficients of the response function for representation of the properties in dependence of the oxalic and nitric acid addition to the sulphuric acid base electrolyte.

Property	Symbol	Unit	a_1	a_2	a_3	a_4	R^2
Porosity	p	%	50	4	13	−14	0.81
Hardness	H_{20}	N/mm^2	2200	900	−1400	1700	0.94
Hardness decline	H^*		1	0.014	−0.04	0.04	0.97

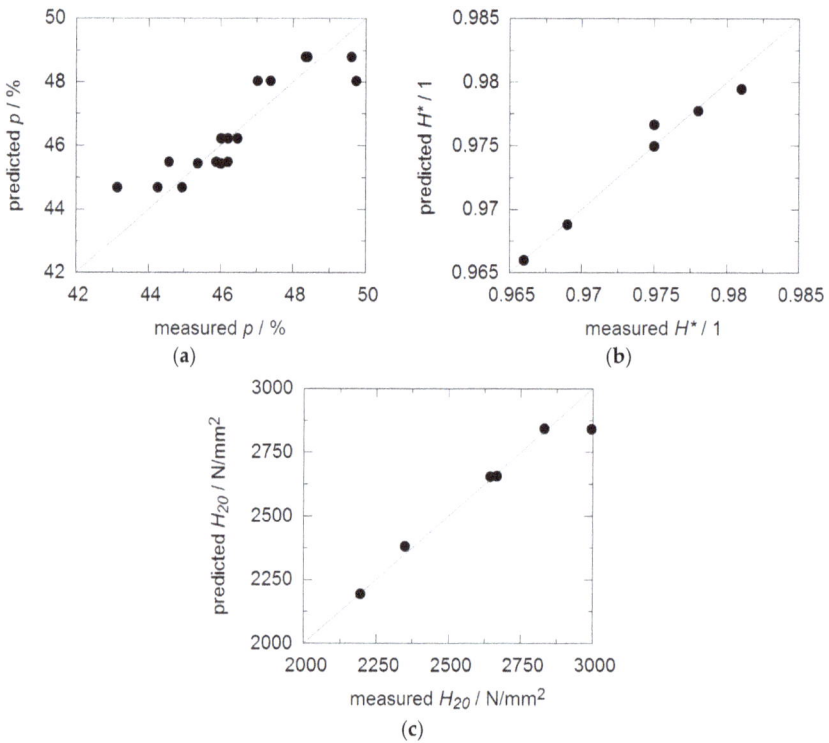

Figure 5. Comparison between values predicted by the response function and measured values for (a) porosity p, (b) hardness coefficient H^*, (c) hardness in a distance of 20 microns from the substrate/oxide interface.

The micron scale porosity of the coatings was examined by electron microscopy using the BSD detector. As can be seen from Figure 6a, the anodic oxide coatings from the base electrolyte contain plenty of spheroidal voids with diameters of less than 5 μm and some irregularly formed voids with dimensions of more than 10 μm. From grey-scale analysis, an average micron scale porosity of 3.4 ± 0.8% was obtained. With the addition of 0.4 mol/L and 0.8 mol/L nitric acid, a similar amount of cracks occurs at large voids (represented by Figure 6b). Hence, the microscale porosity increases to 4.6 ± 1.3% and 4.6 ± 0.8% respectively. With the addition of 0.2 mol/L oxalic acid, the number and volume content of the cracks generally increases for all nitric acid concentrations. The micron scale porosity ranges from 5.8 ± 1.6% for the single addition of 0.2 mol/L oxalic acid (Figure 6c) to 8.4 ± 0.7% for 0.8 mol/L nitric acid and 0.2 mol/L oxalic acid (Figure 6d) respectively. The roughness of the substrate–coating interface close to large voids seems to increase with increasing nitric acid and in particular with increasing oxalic acid concentrations. For the highest additive concentration, several large voids are connected by crack networks (Figure 6d).

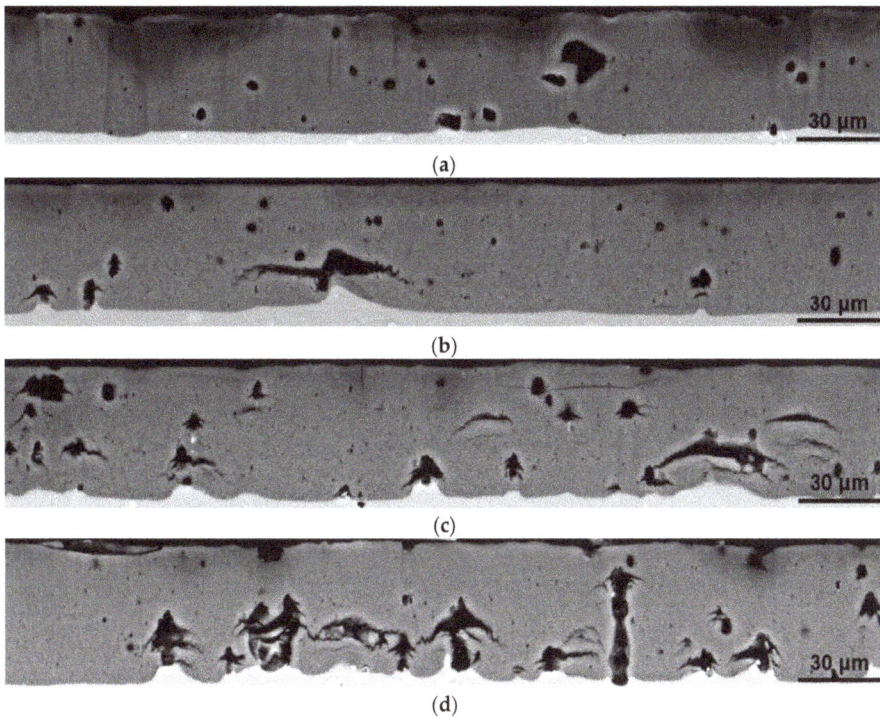

(a)

(b)

(c)

(d)

Figure 6. Backscattered electron detector (BSD) images showing the micron scale porosity of anodic oxide coatings obtained from the following electrolytes: (**a**) 20 vol % H_2SO_4, (**b**) 20 vol % H_2SO_4 + 0.8 mol/L HNO_3, (**c**) 20 vol % H_2SO_4 + 0.2 mol/L $C_2H_2O_4$, (**d**) 20 vol % H_2SO_4 + 0.2 mol/L $C_2H_2O_4$ + 0.8 mol/L HNO_3.

3.3. Coating Adhesion and Abrasion Resistance

In order to evaluate the influence of large micron scale pores and cracks on coating adhesion and coating failure, progressive scratch tests with increasing normal load from 1 to 100 N were performed. Because of the brittleness of the oxide conversion coatings, periodically occurring cracks perpendicular to the scratch direction were already observed from the beginning. As can be seen from the light microscope image Figure 7a, the conversion coatings obtained from the base electrolyte typically chip off after a critical normal force is reached, whereby adhesive failure of oxide plates reaches beyond the scratch. In this case, the critical force of 48.7 ± 2.9 N for the first occurrence of adhesive failure can be determined clearly by both the optical investigation of the scratch and the sudden increase of the remaining scratch depth. A similar failure behavior applies to the single addition of 0.4 mol/L nitric acid, however, the adhesive failure is already observed at a slightly smaller normal force of 45.0 ± 4 N as can be seen from Table 6. For the single addition of 0.2 mol/L oxalic acid, both with and without nitric acid, no spallation of large oxide plates can be observed (Figure 7b). For this reason, it is more difficult to obtain the critical normal force from the optical investigation of the scratch. However, with the aid of the remaining scratch depth curve, a further decrease of the critical normal force to 42.0 ± 5 N and 33.3 ± 2.0 N, respectively, can be derived. For the highest nitric acid concentration of 0.8 mol/L both with and without oxalic acid, the failure mode appears to be very gradually as the remaining scratch depth increases steadily without abrupt increases. The exposure of the bare metallic substrate already occasionally appears at small normal forces due to the abrasive wear of the entire oxide thickness. Hence, a critical normal force for adhesive failure cannot be defined for these samples.

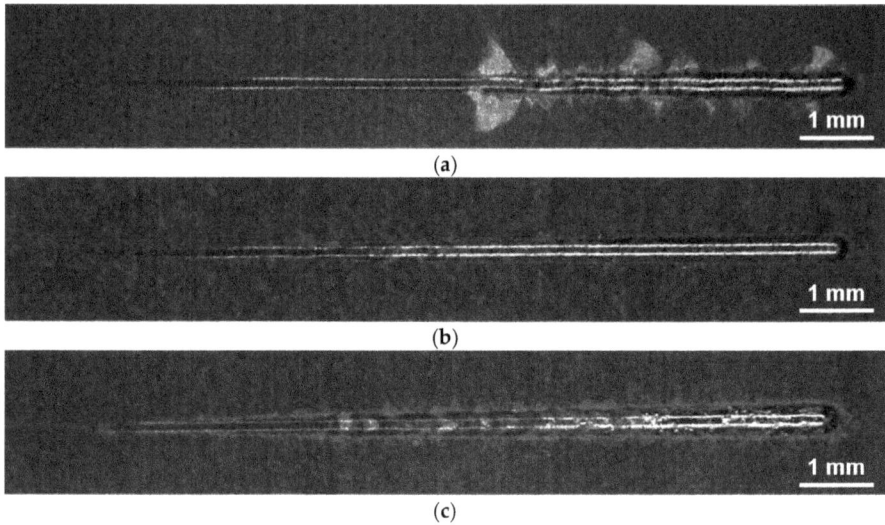

(a)

(b)

(c)

Figure 7. Light microscope images of coating failure after progressive scratch test: (**a**) 20 vol % H_2SO_4, (**b**) 20 vol % H_2SO_4 + 0.2 mol/L $C_2H_2O_4$, (**c**) 20 vol % H_2SO_4 + 0.8 mol/L HNO_3.

Table 6. Survey of scratch properties for all the combinations of additions of oxalic and nitric acid to the sulphuric acid base electrolyte.

Property	Sym-Bol	Unit	Oxalic/Nitric Acid Addition in mol/L					
			0/0	0/0.4	0/0.8	0.2/0	0.2/0.4	0.2/0.8
Critical normal force	F_c	N	48.7 ± 2.9	45.0 ± 4	-	33.3 ± 2.0	42.0 ± 5	-
Scratch energy density	W_R	J/mm^3	1.6 ± 0.2	1.7 ± 0.2	0.9 ± 0.4	1.3 ± 0.1	1.6 ± 0.2	1.0 ± 0.3
Tangential force	F_t	N	0.4 ± 0.0	0.5 ± 0.0	1.4 ± 0.2	0.5 ± 0.0	0.5 ± 0.0	0.9 ± 0.1
Cross-section area	A	μm^2	270 ± 40	309 ± 24	1800 ± 700	347 ± 24	350 ± 50	1100 ± 500

As can be seen from Table 6, the scratch energy density of anodic conversion coatings can be slightly improved through the addition of 0.4 mol/L nitric acid to the base electrolyte. However, this is not due to the reduction of the worn material volume as the cross-section area of the scratches even slightly increases, but due to the slightly increased tangential force. A further increase of the nitric acid concentration impairs the scratch energy density of the coatings considerably. This is due to the significant increase of the worn material volume. Except from the highest nitric acid concentration of 0.8 mol/L, the further addition of 0.2 mol/L oxalic acid to the electrolyte leads to increased cross-section areas and therefore to lower values of the scratch energy density.

4. Discussion

It was shown that both oxalic and nitric acid additions are suitable to improve coating properties. However, solely the addition of nitric acid offers the unique opportunity to enhance the thickness and hardness of anodic oxide coatings and to reduce the electrical energy consumption, simultaneously. This can be attributed to the different effect mechanisms of the additives. It is known that organic additives like oxalates from oxalic acid inhibit the chemical dissolution of alumina at the pore walls in the outer region of anodic conversion coatings [1]. This results in conversion coatings with a higher density and a smaller hardness gradient described by a higher value of the hardness decline H^* in Table 4. Whereas the extended pores of anodic coatings from the base electrolyte allow an easier electrolyte penetration, the accessibility of coatings from electrolytes with oxalic acid addition decreases significantly with increasing coating thickness. Therefore, the electrical resistance increases and a more pronounced rise of the process voltage can be observed for the latter coatings. In contrast to this, the addition of nitric acid allows the reduction of the voltage from the beginning of the process. As already described above, the presence of copper oxide at the interface allows for local oxygen evolution and therefore leads to a reduced current efficiency of oxide growth. One explanation for the beneficial effect of nitric acid may be the accelerated chemical dissolution of the copper oxide at the substrate–electrolyte interface. Aqueous solutions of nitric acid are commonly used to remove the copper enriched black surface layer on copper-rich aluminum alloys after pickling.

When discussing the correlation between the anodizing parameters, porosity, hardness and scratch resistance, it is important to subdivide the porosity in different categories according to their origin: pore channels proceeding orthogonal to the substrate surface, periodically occurring voids along the pore channels due to copper-enrichment and oxygen evolution and microscale voids due to the dissolution of intermetallic phases. Obviously, the chemical dissolution of the pore walls is not substantially reduced by the addition of oxalic acid as the hardness parameters H_{20} and H^* are only slightly enhanced. A reason for this could be the generally low dissolution rate of anodic alumina in 20-vol % sulfuric acid solution at 5 °C. At higher electrolyte temperatures, a stronger effect of the oxalic acid addition has to be expected. In contrast to this, nitric acid addition is suitable to enhance the hardness parameters H_{20} and H^* significantly. Again, this effect can be explained by the accelerated chemical dissolution of copper oxide at the substrate–coating interface. According to this argumentation, the reduction of the oxygen evolution does not only improve the energy efficiency (as already described) but also reduces the amount and volume content of the periodically occurring voids along the pore channels. These results correspond to the results of Morgenstern et al. [8], who recently discovered that thickness and hardness of anodic oxide coatings are improved when the alloying element copper is not homogeneously dispersed in solid solution or in the form of atomic clusters, but concentrated in S-phase (Al$_2$CuMg) precipitates. In this case, the precipitates preferentially dissolve and the detrimental effect of copper is reduced.

The characteristic micron scale voids are developed through the dissolution of micron scale intermetallic phases. Theoretically, the S-phase should completely dissolve during a long-time solution annealing treatment in order to enable the maximum effect of the subsequent age hardening process. Practically, the duration of the solution annealing treatment is limited due to high energy costs and the danger of grain coarsening. For this reason, some S-phase precipitates do not completely dissolve but reshape to a spheroidal form. As already reported in [8,9], these precipitates leave spheroidal voids within the conversion coatings due to their preferential dissolution in the sulphuric acid electrolyte. Because of their limited size of up to 5 µm in diameter and their round shape, they do not act as sharp notches and might stop rather than initiate cracks within the oxide coating. On the other hand, primary phases, e.g., iron- or silicon-rich phases, are precipitated during the solidification of the molten alloy. They are virtually insoluble in the solid aluminum matrix. These precipitates exhibit dimensions of more than 10 µm and an irregular, sharp-edged shape. During anodizing, they convert more slowly than the surrounding aluminum matrix and leave highly porous volumes and flaws within the coating according to [7]. These flaws also exhibit a size of more than 10 µm and sharp edges. Therefore, they might rather act as crack initiation sites. As shown in Figure 6, the susceptibility to cracking increases with increasing nitric acid concentration and especially with the addition of oxalic acid. One reason for this could be the embrittlement of the coatings due to the incorporation of additional elements from the electrolyte. Shih et al. [10] proposed that the enhanced hardness of anodic oxide coatings obtained from a sulphuric acid electrolyte after nitric acid addition results from a higher sulfur content within the oxide. Another explanation could be the influence of nitric acid and oxalic acid on the conversion behavior of the iron-rich intermetallic phases. As can be seen from Figure 6, the roughness of the substrate–coating interface increases in the same order as the number and volume of cracks. The interface roughness results from the different conversion rates of the intermetallic phases and the aluminum matrix. Consequently, it can be argued that the presence of oxalic acid especially inhibits the conversion of the iron-rich phase. Following this argumentation, tensile stresses evolve in the porous oxide ahead of the iron-rich phases as the conversion of the surrounding aluminum matrix is connected with volume expansion. In conjunction with the notch effect of the large, sharp-edged voids, this could finally induce cracking.

The large voids and cracks are significantly larger than the hardness indents. Consequently, the instrumented nanoindentation measurements can only be performed around large pores within more compact oxide volumes so that these voids do not affect the measured hardness values. This is the reason why oxalic acid addition results in both an increasing general porosity and increasing hardness parameters according to Table 4. However, as scratch testing is a more integral characterization method, large voids and cracks influence the coating failure mode and the scratch resistance considerably, as can be seen quantitatively from Table 6 and qualitatively from Figure 7. With an increasing number of large pores and cracks, the failure mode changes from the brittle spallation of oxide plates towards the more gradual coating failure after the abrasion of the entire coating thickness. This is understandable, because compact oxide materials are not able to relieve internal stresses and therefore fail suddenly after reaching a critical stress level. On the other hand, if cracks are already present within the coating, the crack network propagates under normal pressure. Consequently, the indenter can easily remove material volumes, which are completely separated from the surrounding material by the crack network and the oxide coatings are worn more gradually at lower critical normal forces. The scratch energy density is influenced by both the hardness of compact oxide volumes and the micron scale porosity. On the one hand, it is to be expected that the scratch resistance increases with increasing coating hardness. On the other hand, the presence of large pores and cracks deteriorates the abrasion resistance, as already discussed. The optimum scratch energy density can be observed for coatings after the single addition of 0.4 mol/L nitric acid to the base electrolyte as these coatings exhibit both an increased hardness and a comparatively low micron scale porosity.

5. Conclusions

The present work investigates the influences of the single and combined addition of nitric acid and oxalic acid to a sulphuric acid electrolyte on the anodic oxidation behavior of the $AlCu_4Mg_1$ alloy. It was shown that—unlike conventional organic additives—the addition of nitric acid to a sulphuric acid electrolyte enables both the enhancement of coating properties, e.g., hardness by 23%, thickness by 14%, and the reduction of the electrical energy consumption by 5%, simultaneously.

In contrast to this, oxalic acid addition reduces the hardness gradient and slightly increases the hardness of the outer coating regions. Unfortunately, oxalic acid addition is connected with a significant increase in process voltage and therefore an increased energy consumption. The results also suggest that oxalic acid addition decelerates the dissolution of large iron-rich intermetallic phases. This gives rise to internal stresses and causes cracks within the conversion coating.

For the combined addition of nitric and oxalic acid the maximum hardness increase of 36% compared with the base electrolyte and the smallest hardness gradient (represented by the highest values of the hardness H_{20} and the hardness decline H^*) can be achieved. However, the coatings' resistance against the abrasion of a hard counter body (represented by the scratch energy density) generally decreases with increasing additive concentration. Furthermore, the failure mode changes from sudden spallation of oxide plates towards the gradual abrasion of the coating.

By exploiting the different effects of oxalic and nitric acid, the process and coating properties can be optimized with regard to different specifications (e.g., maximum hardness or minimum energy consumption). It is expected that especially the addition of nitric acid is also suitable in order to improve the properties of anodic conversion coatings obtained at ambient temperature, as well. This is the subject of further research.

Acknowledgments: The authors gratefully acknowledge funding by the German Research Foundation (Deutsche Forschungsgemeinschaft, DFG) within the framework of SFB 692 (SFB692B2). The support of Dagmar Dietrich, Dagobert Spieler, Elke Benedix, Christel Pönitz, Paul Clauß and Frank Simchen (all from the Institute of Materials Science and Engineering) is gratefully acknowledged.

Author Contributions: Maximilian Sieber designed and performed most of the anodizing experiments, compiled the response functions, conducted the analysis of variance and wrote the corresponding sections of the paper. Roy Morgenstern performed the scanning electron microscopy investigations, supervised the nanoindentation and scratch tests, analyzed the corresponding results and wrote the corresponding sections of the paper as well as the discussion and conclusion chapters. Ingolf Scharf gave advice to Maximilian Sieber regarding the experimental design, analyzed and discussed the results with Roy Morgenstern and revised the manuscript. Thomas Lampke coordinated the research project. He gave advice to Maximilian Sieber and Roy Morgenstern regarding the focus of the manuscript, the experimental design and appropriate methods. Furthermore, he discussed the results with the other authors and revised the manuscript.

Conflicts of Interest: The authors declare no conflicts of interest. The founding sponsors had no role in the design of the study; in the collection, analyses, or interpretation of data; in the writing of the manuscript, and in the decision to publish the results.

References

1. Giovanardi, R.; Fontanesi, C.; Dallabarba, W. Adsorption of organic compounds at the aluminium oxide/aqueous solution interface during the aluminium anodizing process. *Electrochim. Acta* **2011**, *56*, 3128–3138. [CrossRef]

2. Bensalah, W.; Elleuch, K.; Feki, M.; Wery, M.; Ayedi, H.F. Mechanical and Abrasive Wear Properties of Anodic Oxide Layers Formed on Aluminium. *J. Mater. Sci. Technol.* **2009**, *25*, 508–512.

3. Sieber, M.; Morgenstern, R.; Lampke, T. Anodic oxidation of the AlCu$_4$Mg$_1$ aluminium alloy with dynamic current control. *Surf. Coat. Technol.* **2016**, *302*, 515–522. [CrossRef]

4. Thompson, G.E.; Habazaki, H.; Shimizu, K.; Sakairi, M.; Skeldon, P.; Zhou, X.; Wood, G.C. Anodizing of aluminium alloys. *Aircr. Eng. Aerosp. Technol.* **1999**, *71*, 228–238. [CrossRef]

5. Hashimoto, T.; Zhou, X.; Skeldon, P.; Thompson, G.E. Structure of the Copper–Enriched Layer Introduced by Anodic Oxidation of Copper-Containing Aluminium Alloy. *Electrochim. Acta* **2015**, *179*, 394–401. [CrossRef]

6. Curioni, M.; Roeth, F.; Garcia-Vergara, S.J.; Hashimoto, T.; Skeldon, P.; Thompson, G.E.; Ferguson, J. Enrichment, incorporation and oxidation of copper during anodizing of aluminium-copper alloys. *Surf. Interface Anal.* **2010**, *42*, 234–240. [CrossRef]

7. Ma, Y.; Zhou, X.; Thompson, G.E.; Curioni, M.; Zhong, X.; Koroleva, E.; Skeldon, P.; Thomson, P.; Fowles, M. Discontinuities in the porous anodic film formed on AA2099-T8 aluminium alloy. *Corros. Sci.* **2011**, *53*, 4141–4151. [CrossRef]

8. Morgenstern, R.; Dietrich, D.; Sieber, M.; Lampke, T. Influence of the heat treatment condition of alloy AlCu$_4$Mg$_1$ on the microstructure and properties of anodic oxide layers. *IOP Conf. Ser. Mater. Sci.* **2017**, *181*, 012043. [CrossRef]

9. Morgenstern, R.; Nickel, D.; Dietrich, D.; Scharf, I.; Lampke, T. Anodic Oxidation of AMCs: Influence of Process Parameters on Coating Formation. *Mater. Sci. Forum* **2015**, *825–826*, 636–644. [CrossRef]

10. Shih, H.-H.; Tzou, S.-L. Study of anodic oxidation of aluminum in mixed acid using a pulsed current. *Surf. Coat. Technol.* **2000**, *124*, 278–285. [CrossRef]

metals MDPI

Article

Process Chain for the Production of a Bimetal Component from Mg with a Complete Al Cladding

Wolfgang Förster *, Carolin Binotsch and Birgit Awiszus

Virtual Production Engineering, Chemnitz University of Technology, 09111 Chemnitz, Germany;
carolin.binotsch@mb.tu-chemnitz.de (C.B.); birgit.awiszus@mb.tu-chemnitz.de (B.A.)
* Correspondence: wolfgang.foerster@mb.tu-chemnitz.de; Tel.: +49-371-531-34759

Received: 2 January 2018; Accepted: 19 January 2018; Published: 27 January 2018

Abstract: With respect to its density, magnesium (Mg) has a high potential for lightweight components. Nevertheless, the industrial application of Mg is limited due to, for example, its sensitivity to corrosion. To increase the applicability of Mg, a process chain for the production of a Mg component with a complete aluminum (Al) cladding is presented. Hydrostatic co-extrusion was used to produce bar-shaped rods with a diameter of 20 mm. The bonding between the materials was verified by ultrasonic testing. Specimens with a length of 79 mm were cut off from the rods and forged by using a two-staged process. After the first step (Heading), the Mg core was removed partially by drilling to ensure a complete enclosing of the remaining Mg during the second forging step (Net shape forging). The geometry of the drilling hole and the heading die design were dimensioned with the Finite Element-simulation software FORGE. Hence, a complete Al-enclosed Mg component was achieved by using the described process chain and forming processes. Microstructural investigations confirm the formation of an intermetallic interface as expected.

Keywords: material composite; Al; Mg; co-extrusion; die forging; interface; FEM

1. Introduction

The use of lightweight construction and materials is one way for the automotive industry to limit CO_2 emissions. However, the weight-saving potential of using a single material is limited. Material composites and composite materials can contribute to overcome this challenge by combining two or more different or similar materials. This results in tailored properties and emphasizes the positive properties of the materials and reduces the negative ones. With respect to weight saving, the use of lightweight materials is obvious, and especially the combination of Al and Mg as a material composite. Mg has a low density, but it is sensitive to corrosion. Al alloys (6xxx) provide corrosion resistance and are low-cost lightweight materials. Consequently, a material composite made from Al and Mg should be significantly lighter, depending on the content of Mg, and provide complete protection of the Mg from corrosive media. One approach is the production of a bar-shaped rod by co-extrusion and its use as a semi-finished product for a subsequent forging process. Co-extrusion has been investigated by several authors with different material combinations since the early 1970s [1–5]. Material composites from Mg and Al showed a good formability in contrast to other material combinations, where core fracture was observed [6]. Many investigations have focused on pressing billet geometry, variations of cladding and core material (cladding: Mg, core: Al, and vice versa) [7], or geometry of co-extruded products (e.g., sheets [6], profiles [8], tubes [9] or bar-shaped rods [10–12]). Such studies have shown that an interface between Al and Mg is formed by diffusion during co-extrusion. It comprises two intermetallic phases, Al_3Mg_2 and $Al_{12}Mg_{17}$, and ensures the connection between the materials. Their structure and properties depend on the formation conditions [10–12]. Due to the brittle mechanical properties, the co-extrusion often results in cracks along the interface. The fracture and mechanical properties of the material composites have been investigated in detail by Lehmann [13,14] and Kirbach [15,16]. They found that the interface tensile strength at room temperature was 125 to 145 MPa, and the formability increased at elevated temperatures. The work on the forming of metallic macroscopic composites is very limited. Li et al. produced Al-Mg-Al sheets by hot roll bonding, and investigated the deep drawability [17]. Other investigations have focused on the forging of powder composites [18,19] or metal matrix composites with different particle reinforcement [20,21]. Foydl et al. investigated steel-reinforced aluminum. They used continuous reinforcement only for co-extrusion and discontinuous for forging, and produced a component with different distributions of the wire reinforcement [22]. This was confirmed by Feuerhack [23], who investigated the processing of co-extruded Al-Mg rods by die forging. It was shown that the interface remains intact if compression stresses dominate. Shear stresses caused a fragmentation of the interface, but a new interface was built between the fragments [23].

At the front ends of the bar-shaped rods, the Mg is left unprotected. This problem also remains after subsequent forging. To prevent the Mg from corrosion and increase the durability of the components, it is essential to achieve a complete enclosing by the Al cladding. Therefore, the production of a bimetal component from Mg with a complete Al cladding by using conventional forming processes was investigated.

2. Materials and Methods

For the production and processing of the Al-Mg material composites, commercially available alloys of magnesium (AZ31) and aluminum (AA-6082) were used. Their chemical compositions are given in Table 1. The bar-shaped rods had a diameter of 80 mm (Al) and 60 mm (Mg). For the production of the pressing billets for the co-extrusion process, a blind hole of 58 mm in diameter was drilled and machined into 290 mm-long sections of the Al rods. The Mg rods were also machined to a diameter of 58 mm, and were loose fit into the Al. Figure 1a,b shows a cross-sectional schematic view of the pressing billets with a Mg core and an Al cladding.

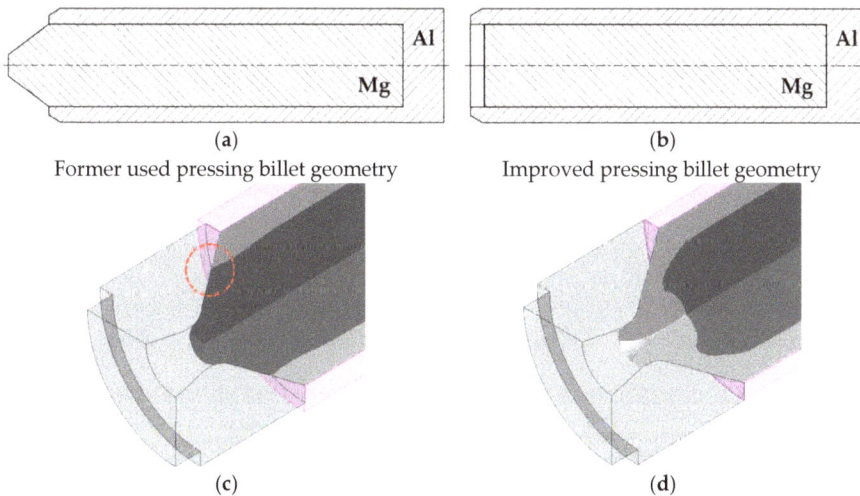

Figure 1. Geometries of the pressing billets and material flow during co-extrusion. (**a**) Longitudinal cross-section of formerly used billet with conical shaped Mg core and (**b**) of the improved billet with shortened Mg core; (**c**) Material flow during infeed of the formerly used billet with contact of the Al-Mg interface on the extrusion die; (**d**) Improved pressing billet geometry ensures that only the Al is in contact with the extrusion die.

Table 1. Chemical compositions of the alloys used for this investigation.

Composition in wt %	Al	Mg	Zn	Mn	Si	Cu	Fe	Ni	Ti	Cr
AA-6082	Balance	0.851	0.082	0.489	0.887	0.107	0.272	-	0.019	0.167
AZ31	2.71	Balance	0.77	0.251	0.017	<0.001	<0.001	<0.002	-	-

2.1. Hydrostatic Co-Extrusion

The hydrostatic co-extrusion process of Al-Mg material composites was extensively investigated by Kittner [24–26]. The container of the press limited the dimensions of the pressing billet to a length of 290 mm and an outer diameter of 80 mm. He varied parameters such as material combinations, the ratio of cladding and core outer diameter, die geometries, and pressing temperatures. The best results were achieved with the material combination AA-6082 (cladding) and AZ31 (core) pressed at a temperature of about 300 °C with an extrusion ratio of 16. Thus, these parameters were used in this investigation. The pressing billet geometry used by Kittner is given in Figure 1a. This geometry with its conical shape leads to contact between the Al-Mg interface and the extrusion die during infeed, as shown in Figure 1c. Thus, lubricant is pressed into the gap between the materials, often preventing bonding. Therefore, the pressing billet geometry was adapted as shown in Figure 1b,d by flattening and shortening the Mg core. The Al protrudes the Mg, and a touching of the contact zone can be prevented.

After the production of the pressing billets, they were heated to extrusion temperature and lubricated before the insertion into the extrusion press. It must be ensured that the lubricant is only applied on the outer surfaces of the cladding material; otherwise, the lubricant may reach the contact zone between the materials and prohibit the bonding between them during co-extrusion. Subsequently, the pressing billets were inserted into the extrusion press and co-extruded with a plunger speed of 3.6 mm/s, as presented schematically in Figure 1c,d. The chamfer at the front end of the pressing billet seals the hydraulic fluid (Ricinus oil). Thus, the pressing pressure can be built up. The hydraulic fluid also acts as a lubricant, by being pressed and hauled between the die and pressing billet. It was found that a mechanical cleaning of the extrusion die from adhering aluminum after each extrusion significantly increases the surface quality of the strands. With the given dimensions of the pressing billet, extruded strands with a length of about 3500 mm and an outer diameter of 20 mm were produced.

2.2. Die Forging

The co-extruded strands were used as semi-finished products for the forging investigations. Basic investigations on the formability of the material composites by die forging were performed by Feuerhack [27,28]. He used three one-stage processes with simple die geometries (as shown in Figure 2) to induce different stress conditions in the components, especially at the interface. The process parameters—including die temperature, specimen temperature and tribology—were investigated regarding their influence on the forging results. He found that best results were achieved with a lubricated die (Gleitmo 820, FUCHS lubritec, Kaiserslautern, Germany) at a temperature of 200 °C and a specimen temperature of 300 °C.

Based on these findings, he developed a more complex component he called SMART-Body. The production of the SMART-Body is a two-stage process comprising heading and net shape forging. This component was also used in the present investigation, and the process steps are given in Figure 3. To achieve a complete filling of the die, the length of the initial billet was increased by 11 mm to 79 mm and cut off from the strand (Figure 3a). The risk of buckling during heading also increased. With an adaptation of the heading die as presented in Section 2.3, this problem could be solved. After cutting off, the billet was heated for half an hour to 350 °C, and the heading die was heated to 200 °C. Subsequently, the billet was forged (Figure 3b). A mechanical press (Raster Zeulenroda, PED 100.3-S4, Zeulenroda, Germany) with a maximum force of 1000 kN was used for the investigation. Thereafter, the Mg core was removed partially by drilling at both front ends of the preform. Variation of the drill hole geometry, drilling depth, and forging temperature was investigated to find the combination for the complete cladding of the Mg during net shape forging. Based on numerical analysis, a drill with a diameter of 15 mm and a drilling angle of 118° was chosen. The drilling depth was 3 mm. Subsequently, the preform was heated up to 350 °C again and forged to net shape with a die temperature of 200 °C (Figure 3c). MoS_2 was used for lubrication.

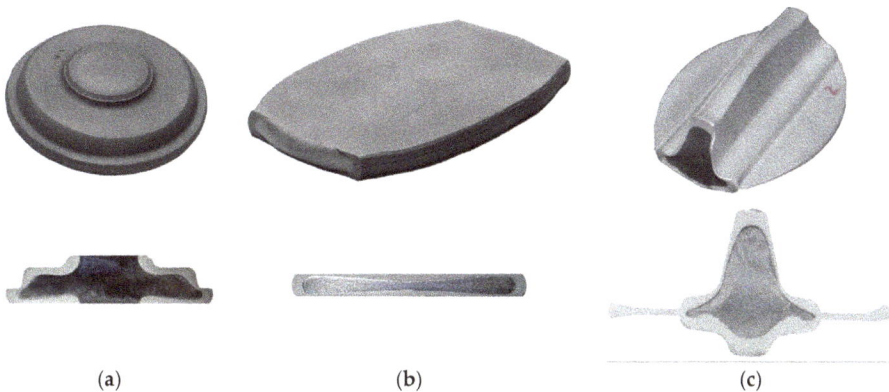

(a)   (b)   (c)

Figure 2. Three simple geometries for the one-staged basic forging investigations. (**a**) Upsetting; (**b**) Spreading; (**c**) Rising [28].

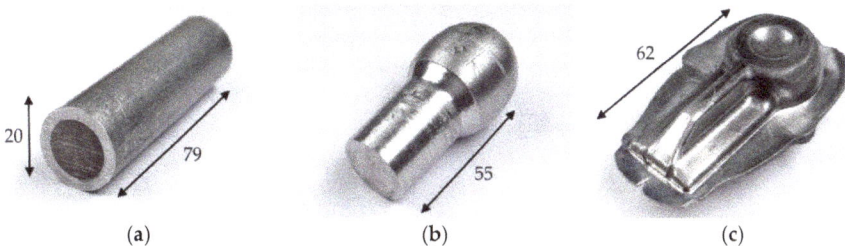

(a)   (b)   (c)

Figure 3. Production stages of the bimetal component with dimensions in mm. (**a**) Initial billet, cut off from a co-extruded rod; (**b**) Preform after first forging step (Heading); (**c**) bimetal component after second forging step (Net shape forging).

2.3. Numerical Analysis of the Forging Process

The development of the process chain was fully accompanied by numerical investigations for parameter identification, process design, and process understanding. A detailed analysis of the co-extrusion process with respect to the prediction of the bonding can be found at KITTNER [26]. Basic investigations on the formability of the material composites by die forging were performed by Feuerhack [28]. He used the simulation system FORGE by Transvalor with implicit time integration. The software was also used for the forging processes in the present investigation based on the process parameters identified by Feuerhack [23,28]. The models can be seen in Figure 4. To limit the computation time, all dies are modeled as rigid bodies, and only the billet was fully deformable. The billet comprises the two bodies cladding (Al) and the core (Mg), connected by a sticking condition, and was thermo-mechanically coupled. The temperature field of the Al and Mg was measured during heating to forging temperature, and no significant differences were found. In accordance with Feuerhack [28], the heat transfer coefficient between the materials was set to 10^6 W·$(m^2 \cdot K)^{-1/2}$ to ensure an equal temperature distribution within the component. The heat transfer between the dies and the component was set to 30×10^3 W·(m²·K)$^{-1/2}$ [28]. In the beginning of the simulation, the die temperature was set to 200 °C and the billet temperature to 350 °C. Tetrahedral elements with a size of 0.6 were used for meshing the billet and 7.6 for the dies. The material data is from the literature [29–31] and Hensel–Spittel equation was used to model the material behavior. It considers the temperature-, strain-, and stress-dependence of materials. The forming history from heading was also considered during net shape forging. Based on experimental investigations, a Tresca friction factor of 0.2 was used to model the lubrication conditions.

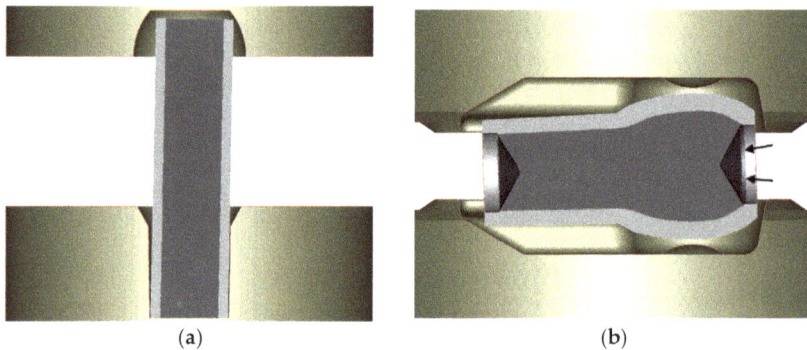

(a)   (b)

Figure 4. Cross-sections of the numerical models for the forging processes. (**a**) Heading die with additional cavity in the upper die, to prevent the billets from buckling; (**b**) Preform after heading with partially-removed Mg core, placed in the net shape forging die; arrows indicate remaining Mg.

Prior to the forging experiments, the heading process was investigated. To ensure a complete die filling, a longer billet was used. Therefore, the model of the former used heading die was adapted to prevent the specimens from buckling. By using a longer billet, the lower die was deepened 9 mm to keep the same size of the head. Additionally, a cavity with the same geometry as the upper part of the head was machined into the upper die. The cavity ensures centering of the billet at the beginning of the heading process, and buckling should be prevented. This was verified with an inclined billet, as can be seen in Figure 4a. The angle between the middle axes of the forging dies and the billet was 1°. Based on these results, the forging dies were also adapted.

After heading, the model of the billet was transferred into a new simulation for the investigation of the removal of the Mg-core and its influence on net shape forging. FORGE provides a special trimming tool to remove material volume by deleting elements. Removing material volume is often problematic in numerical analysis, especially if the model is used in further processing. The elements are connected to forming data (stress, strain, force, etc.), which are also deleted. This missing data can lead to a failing remeshing during transfer from one simulation to the other. In this case, the tool was used to remove the Mg core for the net shape forging after heading with no loss of forming data. Different geometries of the drilling hole (conical, blind), drilling depths (1–4 mm), and forging temperature (300 °C, 350 °C, 400 °C) were investigated and validated with experiments. Figure 4b shows the headed preform with removed Mg core on both sides with a depth of 3 mm and a drilling angle of 118° placed in the net shape forging die.

3. Results

3.1. Hydrostatic Co-Extrusion

Figure 5a–c shows the longitudinal cross-sections of the co-extruded strands. Due to the conical shape of the formerly used pressing billet, there is a section of about 200 mm at the beginning of the strand, where the Mg has no Al cladding (Figure 5a). By contrast, the improved pressing billet geometry results in an enclosed Mg core from the beginning of the process (Figure 5b). The process parameters of the co-extrusion must be adjusted in such a way that an undamaged strand and a sufficient bonding between the materials are ensured. Due to the substantial surface enlargement during co-extrusion, the oxide layers on the materials break up, and the pure materials enter in contact. This results in an accelerated diffusion process and the formation of an interface with two intermetallic phases, Al_3Mg_2 and $Al_{12}Mg_{17}$, as shown in Figure 5d,e.

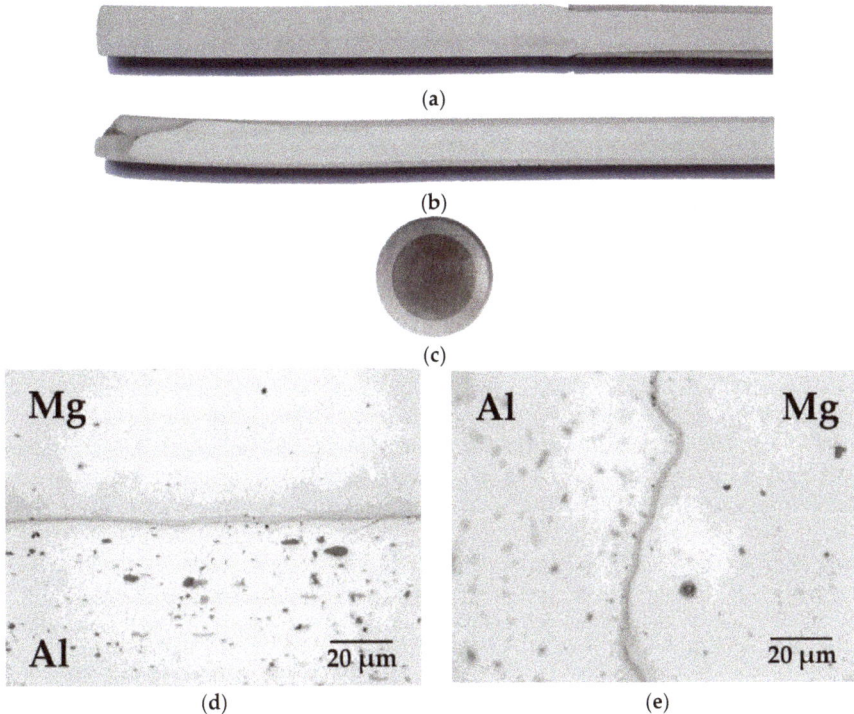

(a)

(b)

(c)

(d)   (e)

Figure 5. Macro- and micrographs from the co-extruded rods. (**a**) Longitudinal cross-section of the beginning of the rod from the formerly used pressing billet with the Mg preceding the Al; (**b**) Improved pressing billet with the Al enclosing the Mg core; (**c**) transversal cross-section of the co-extruded rods. Micrographs of the interface on the (**d**) longitudinal and (**e**) transversal cross-section.

After co-extrusion, the interface had a thickness of 1–2 µm. It is essential to keep the extrusion temperature below the eutectic melting temperatures (436 °C or 450 °C) of the intermetallic phases; otherwise, a eutectic cast microstructure develops at the interface with worse mechanical properties than a diffusion-based microstructure [14,26].

Ultrasonic testing is a straightforward method of investigating whether bonding was achieved. With this non-destructive method, the entire strand can be evaluated by the ultrasonic reflections, as shown in Figure 6. If there is a complete bonding, the ultrasonic waves are reflected at the opposite edge of the strand, resulting in one single strong peak. The two smaller peaks arise from the partial reflection at the interface (Figure 6a). If there is no bonding, the ultrasonic waves are reflected completely and can be identified by many strong peaks corresponding to the repeated reflection at the interface (Figure 6b).

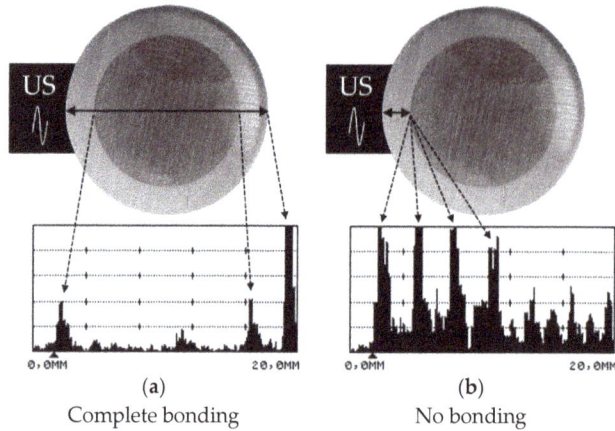

Figure 6. Investigations on the bonding of the co-extruded rods by ultrasonic testing. (**a**) Ultrasonic signals indicating a complete bonding and (**b**) no bonding.

3.2. Die Forging

The investigations on the removal of the Mg core showed that a conical drilling hole results in a complete closing of the front ends, as shown in Figure 7. A complete closing was achieved if the drilling depth was at least 2 mm. An overly-deep drilling hole can lead to the voids remaining between Al and Mg. If the drilling hole is not deep enough, the enclosure is incomplete. The increase of the temperature increases the flash formation by lowering the yield stress. This can promote a flowing of the Mg into the flash due to the increased formability. The interaction of these parameters depends on the component geometry, and must be evaluated for each geometry individually. For the SMART-Body, a complete enclosure of the Mg-core was achieved with a 15 mm conical drilling hole and a depth of 3 mm forged at 350 °C. Depending on the state of closing at the front ends, forging forces of 800 to 1000 kN were measured.

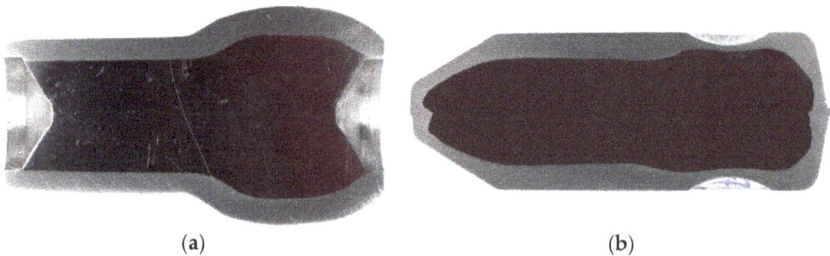

Figure 7. Cross-sections of (**a**) the preform with partially-removed Mg core and (**b**) the net shape forged component.

Figure 8 shows the micrographs of closed front ends and the flash region. They confirm a complete Al cladding. Due to the pressure during forging, the Al is formed around the front ends of the Mg and flows out of the flash gap. The conical Mg core is folded, as indicated by the gap in Figure 8a. A precise positioning of the drilling hole is essential to ensure a complete removal of the Mg at the drilling hole walls. This was the case in the lower part of the component. The interface-free region (dashed arrows) indicates the former drilling hole wall folded onto the Mg. No diffusion was observed between Al and Mg in this region. Due to the compression stress and the missing surface enlargement, the oxide layers (not visible) could not be broken up, and hence prohibit diffusion. In the upper part, the interface (black arrows) was not removed during drilling, and thus it folded onto the Mg. The gap indicates the former drilling hole wall. One part of the interface was partially pressed into the flash gap. Thus, the interface could be pressed out through the flash gap and lower the corrosion resistance by preventing diffusion between the Al. The contact zone of the Al can hardly be identified in Figure 8a. Due to the forming pressure, high strains and temperature influence in that region, and self-diffusion between Al is supposed. No self-diffusion is observed between Mg.

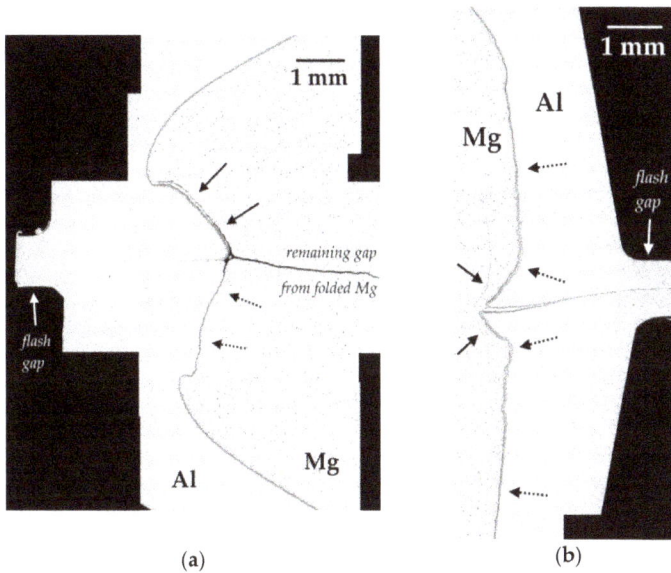

(a)   (b)

Figure 8. Micrographs of the flash gap region at the front ends of the bimetal component.

Due to the different Mg core geometry at the opposite side, it is not possible to remove the Mg sufficiently (also see arrows in Figure 3b). As shown in Figure 8b, the drilling hole walls with the remaining interface were formed onto the Mg, similar to Figure 8a. The contact zone is indicated by black arrows. Together with Mg, the interface was also pressed through the flash gap. Self-diffusion was not observed here, being potentially inhibited by lubricant and other contaminants on the surfaces, which are also pressed through the flash gap. Such impurities must be eliminated to allow diffusion and prevent the contact zone from corrosion. Furthermore, the micrographs show an increase in interface thickness (dashed arrows in Figure 8b). The increase starts from the average value of about 27 μm and reaches values of about 35 μm, and cannot be related to differences in thermal history. This phenomenon was found and described by Feuerhack [28] during upsetting and is confirmed in the present investigation. The hydrostatic pressure during compression of the interface offers a certain formability of the interface and prevents it from fragmentation.

3.3. Numerical Analysis of the Forging Process

The function of the cavity in the upper die is demonstrated in Figure 9a–c for several process steps. After the contact of the inclined billet with the cavity (Figure 9a), it is aligned to the center axis of the forging dies (Figure 9b). Thus, the heading process can be finished without buckling of the billet (Figure 9c). Figure 9d shows the simulation results of the net shape forging of the SMART-Body with a partially-removed Mg core (conical; diameter 15 mm; depth 3 mm). A comparison with the cross-section of the forged component in Figure 7c shows a good agreement between the shape of the Mg core obtained from Finite Element simulation and experiment. Furthermore, the enclosure of the Mg core by the Al is in good agreement with the experimental results. With respect to the chosen boundary conditions, the forging force of about 900 kN was calculated.

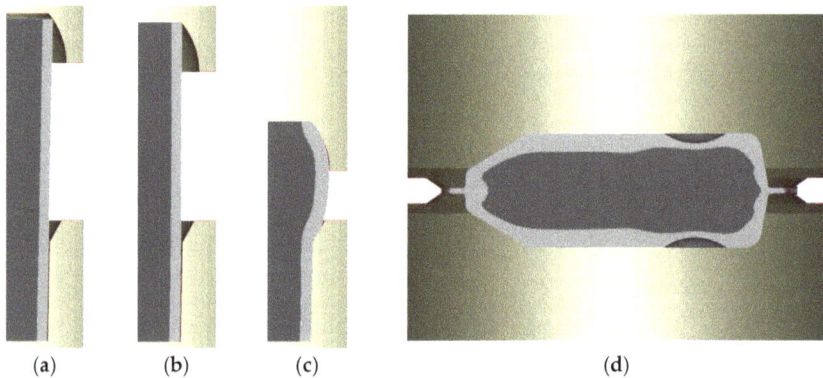

(a)   (b)   (c)   (d)

Figure 9. Cross-sections of the simulation results of the forging processes. (**a–c**) Centering and forming of the inclined billet by the cavity of the upper die; (**d**) Net shape forging of the SMART-Body with partially-removed Mg-core.

4. Conclusions

A process chain for the production of a completely Al-cladded bimetal component with Mg core has been presented. Semi-finished products are produced by hydrostatic co-extrusion at 300 °C with a diameter of 20 mm and a length of about 3500 mm. The bimetal component was produced by die forging in a two-stage process. Billets with a length of 79 mm were cut off from the bar-shaped rods and forged at a temperature of 350 °C. The first process step—heading—is to achieve an appropriate distribution of material volume before net shape forging. To prevent the billets from buckling, the heading die was adapted with an additional cavity in the upper die. The cavity ensures centering of the billets in the first process steps. After heading, the Mg core was partially removed by drilling at the front ends. The remaining and protruding Al encloses the Mg core during the second forging step. The contact zone of the enclosing Al is located in the center of the flash gap and self-diffusion is supposed between the Al-cladding at the front ends.

The described process chain shows one approach to producing lightweight components from a metallic material composite. However, the size of such components is limited with respect to the size of the co-extruded rods. Nonetheless, there is strong potential in this field for future investigations. The focus might be on the improvement of this process chain or the development of other strategies and material combinations.

Acknowledgments: The authors are grateful to the German Research Foundation for their financial support of these investigations.

Author Contributions: W. Förster was responsible for performing the experimental and numerical investigations and wrote this paper. All authors have discussed the results.

Conflicts of Interest: The authors declare no conflict of interest.

References

1. Osakada, K.; Limb, M.; Mellor, P.B. Hydrostatic extrusion of composite rods with hard cores. *Int. J. Mech. Sci.* **1973**, *15*, 291–307. [CrossRef]
2. Story, J.M.; Avitzur, B.; Hahn, W.C. The effect of receiver pressure on the observed flow pattern in the hydrostatic extrusion of bimetal rods. *ASME J. Eng. Ind.* **1976**, *98*, 909–913. [CrossRef]

3. Avitzur, B.; Wu, R.; Talbert, S.; Chou, Y.T. An analytical approach to the problem of core fracture during extrusion of bimetal rods. *ASME J. Eng. Ind.* **1985**, *107*, 247–253. [CrossRef]
4. Kleiner, M.; Schomäcker, M.; Schikorra, M.; Klaus, A. Manufacture of extruded and continuously reinforced aluminum profiles for ultra-lightweight constructions. *Mater. Werkst.* **2004**, *35*, 431–439. [CrossRef]
5. Jang, D.H.; Hwang, B.B. Deformation Analysis of Co-Extrusion Process of Aluminum Alloy and Copper Alloy. *Key Eng. Mater.* **2007**, *340*, 645–648. [CrossRef]
6. Engelhardt, M.; Grittner, N.; Haverkamp, H.; Reimche, W.; Bormann, D.; Bach, F.-W. Extrusion of hybrid sheet metals. *J. Mater. Process. Technol.* **2012**, *212*, 1030–1038. [CrossRef]
7. Feng, B.; Xin, Y.; Hong, R.; Yu, H.; Wu, Y.; Liu, Q. The effect of architecture on the mechanical properties of Mg–3Al–1Zn rods containing hard Al alloy cores. *Scr. Mater.* **2015**, *98*, 56–59. [CrossRef]
8. Muehlhause, J.; Gall, S.; Mueller, S. Simulation of the co-extrusion of hybrid Mg/Al profiles. *Key Eng. Mater.* **2010**, *424*, 113–119. [CrossRef]
9. Golovko, O.; Bieliaiev, S.M.; Nürnberger, F.; Danchenko, V.M. Extrusion of the bimetallic aluminum-magnesium rods and tubes. *Forsch. Ingenieurwesen* **2015**, *79*, 17–27. [CrossRef]
10. Negendank, M.; Mueller, S.; Reimers, W. Coextrusion of Mg–Al macro composites. *J. Mater. Process. Technol.* **2012**, *212*, 1954–1962. [CrossRef]
11. Priel, E.; Ungarish, Z.; Navi, N.U. Co-extrusion of a Mg/Al composite billet: A computational study validated by experiments. *J. Mater. Process. Technol.* **2016**, *236*, 103–113. [CrossRef]
12. Paramsothy, M.; Srikanth, N.; Gupta, M. Solidification processed Mg/Al bimetal macrocomposite: Microstructure and mechanical properties. *J. Alloys Compd.* **2008**, *461*, 200–208. [CrossRef]
13. Lehmann, T.; Stockmann, M.; Kittner, K.; Binotsch, C.; Awiszus, B. Fracture mechanical properties of Al/Mg compounds and yield behavior of the material during the production process. *Mater. Werkst.* **2011**, *42*, 612–623. [CrossRef]
14. Lehmann, T. Experimentell-numerische Analyse Mechanischer Eigenschaften von Aluminium/Magnesium-Werkstoffverbunden. Ph.D. Thesis, Chemnitz University of Technology, Chemnitz, Germany, 29 June 2012.
15. Kirbach, C.; Lehmann, T.; Stockmann, M.; Ihlemann, J. Digital image correlation used for experimental investigations of Al/Mg compounds. *Strain* **2015**, *51*, 223–234. [CrossRef]
16. Kirbach, C.; Stockmann, M.; Ihlemann, J. Rate dependency of interface fragmentation in Al-mg-compounds. In Proceedings of the Conference Abstract of the 34th Danubia-Adria Symposium on Advances in Experimental Mechanics, Trieste, Italy, 19–22 September 2017; pp. 167–168.
17. Li, C.; Chi, C.; Lin, P.; Zhang, H.; Liang, W. Deformation behavior and interface microstructure evolution of Al/Mg/Al multilayer composite sheets during deep drawing. *Mater. Des.* **2015**, *77*, 15–24. [CrossRef]
18. Behrens, B.-A.; Kosch, K.-G.; Frischkorn, C.; Vahed, N.; Huskic, A. Compound forging of hybrid powder-solid-parts made of steel and aluminum. *Key Eng. Mater.* **2012**, *504*, 175–180. [CrossRef]
19. Shishkina, Y.A.; Baglyuk, G.A.; Kurikhin, V.S.; Verbylo, D.G. Effect of the deformation scheme on the structure and properties of hot-forged aluminum-matrix composites. *Powder Metall. Met. Ceram.* **2016**, *55*, 5–11. [CrossRef]
20. Liao, W.; Ye, B.; Zhang, L.; Zhou, H.; Guo, W.; Wang, Q.; Li, W. Microstructure evolution and mechanical properties of SiC nanoparticles reinforced magnesium matrix composite processed by cyclic closed-die forging. *Mater. Sci. Eng. A* **2015**, *642*, 49–56. [CrossRef]
21. Purohit, R.; Qureshi, M.M.U.; Kumar, B. Effect of Forging on Aluminum Matrix Nano Composites: A Review. *Mater. Today Proc.* **2017**, *4*, 5357–5360. [CrossRef]
22. Foydl, A.; Pfeiffer, I.; Kammler, M.; Pietzka, D.; Matthias, T.; Jäger, A.; Tekkaya, A.E.; Behrens, B.-A. Manufacturing of Steel-reinforced Aluminum Products by Combining Hot Extrusion and Closed-Die Forging. *Key Eng. Mater.* **2012**, *504*, 481–486. [CrossRef]
23. Feuerhack, A.; Binotsch, C.; Awiszus, A. Formability of hybrid aluminum-magnesium compounds. *Key Eng. Mater.* **2013**, *554*, 21–28. [CrossRef]
24. Kittner, K.; Awiszus, B. Numerical and experimental investigations of the production processes of coextruded Al/Mg-compounds and the strength of the interface. *Key Eng. Mater.* **2019**, *424*, 129–135. [CrossRef]
25. Kittner, K.; Awiszus, B. The process of co-extrusion—An analysis. *Key Eng. Mater.* **2012**, *491*, 81–88. [CrossRef]
26. Kittner, K. Integrativer Modellansatz bei der Co-Extrusion von Aluminium-Magnesium-Werkstoff-Verbunden. Ph.D. Thesis, Chemnitz University of Technology, Chemnitz, Germany, 11 May 2012.

27. Binotsch, C.; Nickel, D.; Feuerhack, A.; Awiszus, B. Forging of Al-Mg compounds and characterization of interface. *Procedia Eng.* **2014**, *81*, 540–545. [CrossRef]
28. Feuerhack, A. Experimentelle und Numerische Untersuchungen von Al-Mg-Verbunden Mittels Verbundschmieden. Ph.D. Thesis, Chemnitz University of Technology, Chemnitz, Germany, 23 May 2014.
29. Kammer, C. *Aluminium-Taschenbuch*, 16th ed.; Aluminium-Verlag: Düsseldorf, Germany, 2000; ISBN 3870172754.
30. Kammer, C. *Magnesium-Taschenbuch*, 1st ed.; Aluminium-Verlag: Düsseldorf, Germany, 2000; ISBN 3870172649.
31. Doege, E.; Janssen, S.; Wieser, J. Characteristic values for the forming of the magnesium alloy AZ31. *Mater. Werkst.* **2001**, *32*, 48–51. [CrossRef]

![metals logo] *metals*

MDPI

Article

A Fragmentation Criterion for the Interface of a Hydrostatic Extruded Al-Mg-Compound

Carola Kirbach *, Martin Stockmann and Jörn Ihlemann

Institute of Mechanics and Thermodynamics, Chemnitz University of Technology, D-09126 Chemnitz, Germany; martin.stockmann@mb.tu-chemnitz.de (M.S.); joern.ihlemann@mb.tu-chemnitz.de (J.I.)
* Correspondence: carola.kirbach@mb.tu-chemnitz.de; Tel.: +49-371-53139612

Received: 15 December 2017 ; Accepted: 21 February 2018; Published: 2 March 2018

Abstract: Due to the higher demand for energy efficient products, light-weight constructions have become more important in recent years. An innovative, hydrostatic extruded Al-Mg-compound used here combines the corrosion resistance of aluminium with the outstanding lightweight properties of magnesium. During the production process, a thin boundary layer is built between the two basic materials. Investigations on further hot forming processing revealed a good formability of these compounds despite the fact that the boundary layer splits into fragments during forging and a new secondary boundary layer is built when the basic materials between the fragments come into contact again during the continuous deformation. The aim of the research is now to investigate fragmentation depending on the deformation rate and boundary layer thickness, which increases during the heat-up process in preparation of forging. For this purpose, a channel compression test is used in conjunction with a special newly developed specimen shape. The metallographic evaluation of the boundary layer reveals a strong dependency of fragmentation on the deformation rate and the boundary layer thickness. With the aid of a numerical simulation, an individual critical stretch could be determined at which fragmentation starts, and provide guidance for an optimal forging process design.

Keywords: aluminium; magnesium; hydrostatic extrusion; compound; channel compression test; microstructure; fragmentation

---

1. Introduction

Within the Collaborative Research Center 692, the potential of aluminium-based light-weight materials is utilized under the consideration of many influencing factors during processing. In one field, the research focuses on Al-based hybrid structures that combine magnesium and aluminium. Such an Al-Mg-compound connects the corrosion resistance of aluminium with the outstanding lightweight properties of magnesium, having a 35% lower density. This offers the potential to comply with the requirement of weight saving in several fields of industry, e.g., automotive [1] and aeronautical. Therefore, several investigations have been done since the early 1970s on the co-extrusion process with different material combinations [2–4].

Through the production process of these compounds, a 1–2 μm thin interface develops everywhere between the two basic materials. The compound together with the interface was under examination regarding production process and resulting bonding quality, strength of the basic materials, interfacial strength, residual stresses, fracture mechanical properties and formability [5–9] as described below. The formability of such hybrid structures including their interface are rarely investigated. There are only publications from Kosch [10,11], Foydl [12] and Feuerhack [5], who made two main findings during further forging processes, evaluating the formability of the compounds. Firstly, the heat-up process in preparation of forging leads to an enormous growth of the boundary layer up to a thickness of at least 25 μm. Secondly, the boundary layer splits into fragments during forging and a new thin secondary boundary layer is built when the basic materials between the fragments come into contact again during continuous deformation. Continuing these investigations, the fragmentation process is analysed closer in this paper regarding the dependency on the deformation rate and boundary layer thickness due its importance for the bearing capacity of the whole hybrid structure.

The compound presented here is produced by hydrostatic co-extrusion at CEP GmbH, Freiberg, Germany. The principle production process is shown in Figure 1. Hydrostatic co-extrusion is characterized by a force transmission through a pressure medium that surrounds the bi-material bolt and leads to good lubrication between bolt and container or die during pressing. The materials used were the lightweight alloys AZ31 and AA6082. The outer diameter of the manufactured compound amounts to 20 mm and the inner diameter to 14.5 mm. Further information about the process and optimisation possibilities can be found in [5,7,13,14].

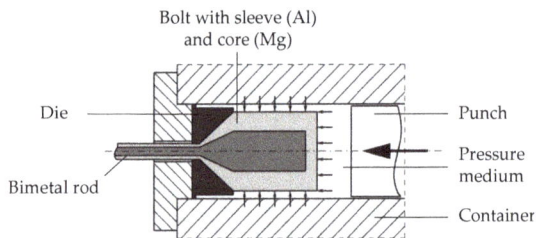

Figure 1. Principle of hydrostatic co-extrusion.

After the manufacturing process at 300 °C, the compound cools down to room temperature causing residual stresses due to differences in the thermal expansion coefficient. Nevertheless, the residual stresses stay at a low level of 16 N/mm^2 [6,15,16] and do not have to be considered during stress analysis in subsequent loading tests.

Investigations with bending tests regarding the determination of the interfacial strength were performed in a wide temperature range from room temperature up to 400 °C [6,17–19]. The loading condition and the specimen shape are designed such that failure occurs under a tensile normal stress. At room temperature, the bending specimens failed with a brittle fracture and revealed a high interfacial strength of about 140–250 N/mm^2. The brittle material behavior of the boundary layer under bending at room temperature correlates with the results of the fracture mechanical tests [6,20]. An increase of the temperature leads to a massive reduction of the interfacial strength (400 °C: 10–60 N/mm^2). Furthermore, a good formability of the compound at high temperatures could be shown, which resulted partly in failure through large plastic deformation under bending. In these cases, the bending strength of the boundary layer is higher than the yield stress of the basic materials, which is important for further forging processes [21].

The rotationally symmetric compound has be further processed by forging, utilizing it for various purposes. Investigations regarding the forging processes upsetting, spreading and rising showed a good formability of the compound at 300 °C [5,7,9]. During the heat-up process in preparation of forging, the boundary layer grows up to a thickness of at least 25 μm due to diffusion based processes [22]. Depending on the load direction during forging, the boundary layer stays intact or splits into fragments. The reason can be found in the elongated bar-shaped microstructure (Figure 2a,b). In the case of a load in the transverse direction of the grains or under a hydrostatic pressure, the boundary layer exhibits a high formability. A load in the direction of the grains leads to a fragmentation following a damage mechanism. At first, the boundary layer splits (step 1). After fracture, the fragments drift apart and, depending on the further deformation, potentially rotate (step 2). In the case of a new contact of the two basic materials between the fragments, a new secondary boundary layer is built as shown in Figure 2c (step 3). The newly developed damage model from Feuerhack already enables determining critical areas regarding fragmentation. The missing differentiation between areas with stretching and compression makes a critical check for plausibility necessary. Due to the also missing influence of the grain structure and specific material properties, the damage model cannot show a load-dependent deformability of the boundary layer either. The quantification of the critical stretch regarding the onset of the fragmentation is part of the presented work and continues the work of Feuerhack.

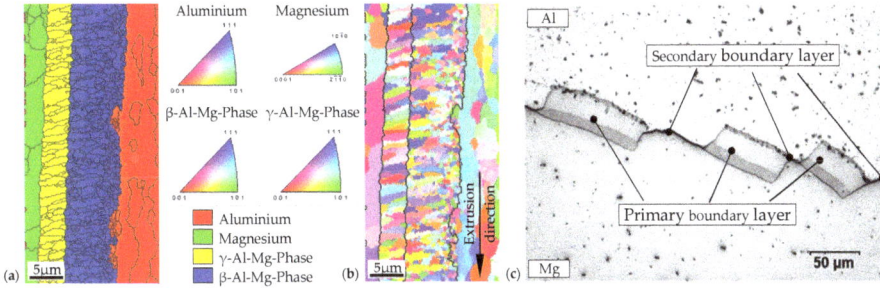

Figure 2. Representation of the grain structure (**a**) Electron backscatter diffraction phase map; (**b**) Inverse pole figure map; (**c**) principle state of the boundary layer after fragmentation [5].

Quantifying the critical stretch regarding the onset of fragmentation requires an experiment that causes a stretch of the boundary layer. In addition, different load states have to be investigated to detect the starting time of fragmentation. Furthermore, the metallographic analysis of the fragmentation state is only possible in a cutting plane. Therefore, the fragmentation state should be homogeneous in one direction, making the evaluation independent from the position of the normal cutting plane along that direction.

Tensile tests cause a stretch of the boundary layer under a known stress state. Due to a limited degree of deformation and the necessary amount of tests to cover different load states, tensile tests are not suitable for this purpose.

The channel compression test [23] together with a newly developed specimen shape causes an unequal distribution of the boundary layer stretch. Such a test is very well-suited due to the possibility to examine different deformation states of the boundary layer in one specimen. As the name of the channel compression test suggests, the deformation is limited to a compression of the height and a stretching along the channel. The sample width stays steady under ideal circumstances, leading to a plain strain state, which is another main advantage of such a test.

2. Materials and Methods

The lateral sample guiding is realised by two steering plates screwed onto a base plate (Figure 3a,b). By the through bores, they can be adjusted to the present stamp position and the width of the actual specimen. To prevent sliding of the steering plates during compression, additional clamping jaws are used. All areas that are in contact with the specimen are polished and lubricated with a MoS$_2$-paste. The loading device can be implemented into different testing machines together with a temperature chamber. All compression tests performed with a stamp velocity of 2 mm/s are realised in an 100 kN ZWICK/ROELL universal testing machine (Zwick GmbH & Co. KG, Ulm, Germany) with an additional 5 kN force transducer. For all other tests, a 50 kN INSTRON hydraulic testing machine (Instron GmbH, Darmstadt, Germany) is used (Figure 3c). The air heater is controlled by the specimen temperature that is about 300 °C during testing. The thermal isolation of the loading device is attained by ceramic punches, serving also as an universal base for different loading devices and connecting parts for different testing machines.

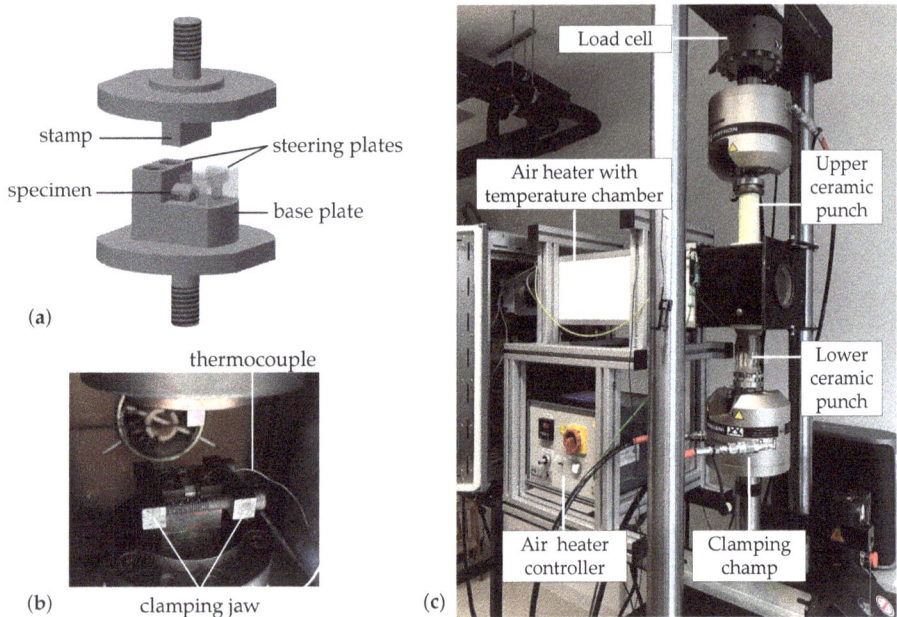

Figure 3. Loading device for the channel compression test (**a**) Computer-aided design depiction; (**b**) experiment; (**c**) implementation into a temperature chamber and a 50 kN INSTRON hydraulic testing machine.

The dimensions of the specimen used for the channel compression test are shown in Figure 4 together with the associated coordinate system. The reduction of the height (y-direction) during testing correlates with a radial compression, taking the point of withdrawal in the compound into account. The specimen elongation parallel to the channel is equivalent to an elongation in the x-direction of the specimen and respectively the tangential direction in the compound. A deformation in the z-direction of the specimen that conforms to the extrusion direction is negligible through the steering plates.

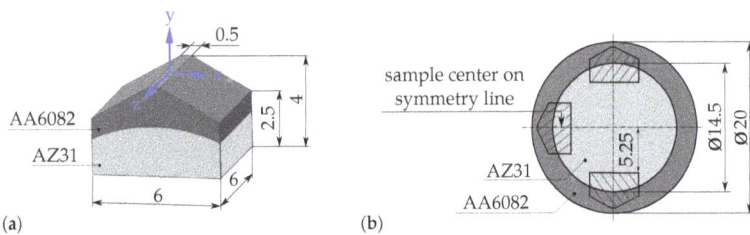

Figure 4. Channel compression specimen (all units are in mm) (**a**) dimensions and coordinate system; (**b**) positions in the undeformed cross-section.

All channel compression tests are summarized in Table 1. Every set of parameters is represented by three specimens extracted from the same rope section. The boundary layer thickness can be adjusted by a previous thermal treatment as shown in [22]. This growth is based on the diffusion of atoms across the interface at a temperature below the eutectic temperature. The lowest thickness of 7.5 µm emerges already through the loading device heat-up process to 300 °C, lasting 120 min, while the specimen is already located in the channel. The specimen temperature complies with the billet temperature during the former hot mass forming processes carried out by Feuerhack [5].

Boundary layer thicknesses of 10 μm and 12.5 μm are reached by an additional dwell time before compression for 15 min and 30 min respectively. To achieve a thickness of 25 μm without a longer dwell time, the specimen is heated up to a temperature of 350 °C for 10 min and then quickly cooled down to test temperature of 300 °C by opening the oven door.

Table 1. Overview of all parameter sets.

Boundary Layer Thickness d	Stamp Velocity	Abbreviation
25 μm	0.2 mm/s	V1
25 μm	2 mm/s	V2/D4
25 μm	20 mm/s	V3
25 μm	200 mm/s	V4
7.5 μm	2 mm/s	D1
10 μm	2 mm/s	D2
12.5 μm	2 mm/s	D3

To ensure the, in some cases, very high stamp velocities from the beginning of the compression, a run-up is included in the test procedure. A run-up distance of 1 mm is sufficient to reach the required stamp velocities. Due to the elasticity of the experimental setup, the traverse stroke differs from the sample height reduction and has to be adjusted to ensure a compression of 1 mm. Immediately after the compression, the sample is removed from the channel and cools down to room temperature. This step prevents a subsequent thermal treatment and accordingly an additional boundary layer growth.

2.1. Metallographic Evaluation

At first, the check of the real boundary layer thickness is performed. With the aid of the digital light microscope Leica VZ700 C with a DVM2500 camera (Leica Microsystems GmbH, Wetzlar, Germany), the sample edges are closely investigated. These positions are appropriate due to the negligible deformation during the compression because of the special sample shape (see Section 3.2). The boundary layer thickness here corresponds to the state directly before the compression test. To minimize the influence of local fluctuations in the thickness and for a statistical coverage, five measurement points at both outer edges are analyzed per sample. The present average thicknesses differ only slightly from the desired ones. The maximum deviations (2.5 μm) and fluctuations can be detected for a thickness of 25 μm, resulting from distinctions in the heating process.

In regards to the following evaluation of the boundary layer structure with respect to the number and length of fragments together with the distance between them, it is worthwhile to depict the whole boundary layer in one image. The related microscope software enables an image extension to cover large structures with a sufficient resolution. After preparing boundary layer images of all specimens, each fragment and gap is sized (Figure 5) to receive information about the boundary layer behavior under load. If the basic materials come into contact again between two fragments, a new secondary boundary layer is built that still belongs to the gap and is not treated as a new fragment. The deformation behavior of the secondary boundary layer cannot be investigated with the channel compression test shown here. Due to an unknown point of development, this layer is only exposed to an unknown part of the complete deformation. Furthermore, the thickness differs enormously from the primary boundary layer, resulting probably in another deformation behavior as the following investigations indicate. In the following assessment, the designation interface includes the complete contact area of the two basic materials and therefore contains fragments and gaps. The designation boundary layer only contains the fragments.

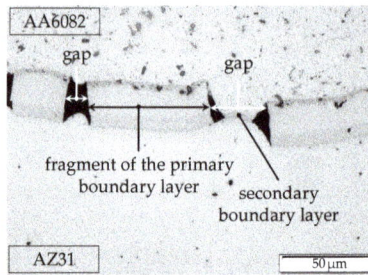

Figure 5. Example image of the boundary layer after channel compression with fragments and gaps.

The schematic sketch in Figure 6 shows the measured dimensions above described together with the associated parameters used for the evaluation. The length l_0 corresponds to the theoretical initial boundary layer length in an ideal manufactured specimen without the meandering shape of the real boundary layer. A measurement of the real initial boundary layer is not possible due to the necessary new surface preparation after channel compression, resulting in a material removal and accordingly another cross section. Additionally, the disposal of the embedding resin after measuring l_0 could damage the specimen and its boundary layer.

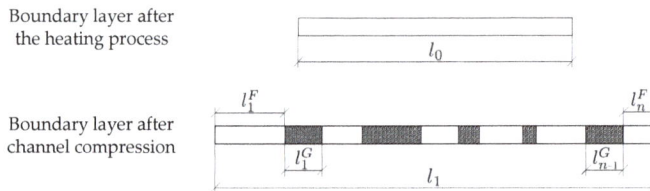

Figure 6. Schematic sketch of the boundary layer before and after channel compression.

Receiving initial information about the ductility of the boundary layer, the length of the fragments is summed up to l_T^F in Equation (1) to compute the average fragment strain according to Equation (2). ε^F should not be seen as a breaking elongation of the boundary layer due to the strong strain inhomogeneity in the specimen due to the special specimen shape:

$$l_T^F = \sum_{i=1}^{n} l_i^F,$$

(1)

$$\varepsilon^F = \frac{l_T^F - l_0}{l_0}.$$

(2)

The lengths of the gaps are also summed up to l_T^G in Equation (3) and used to calculate the percentage of gaps on the interface p^G after channel compression according to Equation (4). In contrast to ε^F, the variable p^G is only based on dimensions measured at the deformed cross section and is not subjected to any assumptions like l_0:

$$l_T^G = \sum_{i=1}^{n-1} l_i^G,$$

(3)

$$p^G = \frac{l_T^G}{l_1} \times 100\%.$$

(4)

2.2. Stretch of the Boundary Layer

The special specimen shape induces an irregular distribution of the boundary layer stretch. This results in an area of fragmentation that is always located in the middle of the interface. In the outer parts, no fragmentation occurs. To determine the distribution of the boundary layer stretch, a numerical simulation of the channel compression test is performed with ABAQUS 6.14-4 (Dassault Systèmes Simulia Corp., Providence, RI, USA).

To verify the numerical simulation, the profile of the boundary layer after the compression is used. A comparison of all 21 boundary layer profiles reveals no significant differences independent of the fragmentation state. The measurement of the profile is done by a dot-wise scanning in the image of the whole specimen shown. The specimen 3 of the parameter set V3 serves as a reference. Due to the approximately identical profiles, the deformation behavior and the fragmentation state of the boundary layer respectively seem to have no influence on the specimens macroscopic deformation. For that reason, the boundary layer is not included in the simulation.

The three-dimensional simulation is based on some simplifications and makes use of both symmetries. A schematic display of the numerical setup is shown in Figure 7 together with the used mesh geometry. The stamp, steering plates and base plate consist of rigid bodies. The steering plates and the base plate are connected to the channel and rigidly clamped. The stamp performs a vertical shift of 1 mm. Shifts in all other directions and rotations are suppressed. The determined stamp shift does not correlate with the traverse stroke in the experiments due to the missing elasticities of the experimental setup in the numerical simulation. The stamp velocity is 200 mm/s. At the sections with contact boundary conditions, friction is assumed. Between channel and specimen, a friction coefficient of 0.1 is determined due to the lubrication. The specimen's top and the stamp are not lubricated, which is why the friction coefficient here is increased to 0.2. The heat-up and cool-down processes as well as the load relief after the compression test are not considered in the simulation.

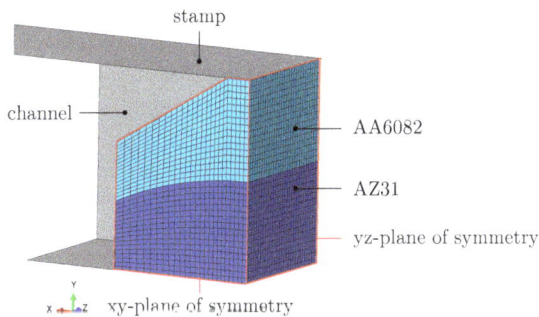

Figure 7. Numerical simulation of the channel compression test - schematic display of the setup together with the used mesh geometry.

The material behavior in the simulation is based on flow curves determined by tensile tests with dwell times and monolithic specimens extracted from the hydrostatic extruded rode. More information about the tensile tests can be found in [24]. A tension–compression–anisotropy that occurs through the production process is not considered in the simulation. According to NOSTER [25], the influence of the stress direction on the deformation behavior of hot-rolled AZ31 at room temperature disappears at 300 °C. Therefore, the use of an isotropic material behavior is permitted.

The implementation of rate-dependent flow curves in ABAQUS is possible through the specification of a quasi-static flow curve together with the increase of the yield stress in accordance with the equivalent plastic strain rate. The flow curves for AZ31 and AA6082, respectively, are described in Figures 8 and 9. The approximate quasi-static flow curves correspond to the experimentally determined curves at a logarithmic deformation rate of $5 \times 10^{-5}\,\text{s}^{-1}$. The deformation rate in conjunction with a stamp velocity of 200 mm/s exceeds the deformation rate of the tensile tests, for which reason the flow curves are extrapolated.

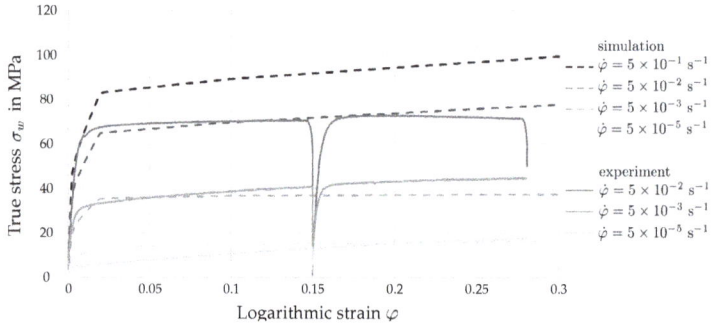

Figure 8. Numerical and experimental flow curves of AZ31.

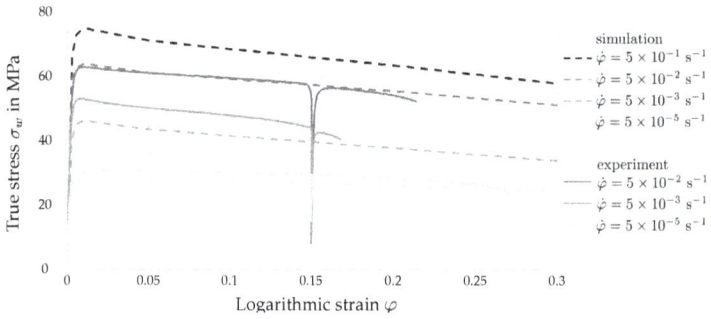

Figure 9. Numerical and experimental flow curves of AA6082.

Both the quasi-static flow curves as well as the increasing of the yield stress are adapted and extrapolated, respectively, with the aim of fitting the boundary layer profile in the simulation to the measured one. The deviations from the measured flow curves that could emerge are subordinated to this aim. A comparison of the real boundary layer profile with the simulation revealed a satisfactory accordance (Figure 10), allowing to deduce the stretch of the boundary layer from the numerical simulation.

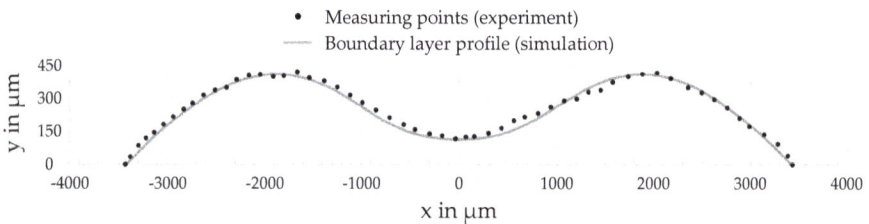

Figure 10. Comparison of the measured boundary layer profile (parameter set V3 specimen 3) and the profile based on the numerical simulation (mirrored at $x = 0$), no true-to-scale illustration.

3. Results and Discussion

3.1. Metallographic Evaluation

Receiving initial information about the ductility of the boundary layer, the average fragment strain and the percentage of gaps are evaluated. The results regarding the dependency on stamp velocity of a 25 µm thick boundary layer are shown in Figure 11. The large deviations between samples with the same boundary layer thickness result from irregularities in the microstructure (e.g., inclusions, the meander shape of the boundary layer, etc.) and small deviations during testing (e.g., testing temperature, stamp stroke, etc.). The maximum average fragment strain of ε^F = 6.9% is reached with a stamp velocity of 0.2 mm/s. Such a deformation is very slow in comparison to mass forming processes and not industrially used. The remaining higher stamp velocities lead to no significant fragment strain ε^F and thereby reveal a brittle material behavior. The shift from brittle to ductile material behavior lies between the stamp velocities 0.2 mm/s (ε^F = 6.9%) and 2 mm/s (ε^F = 1.9%). Such a brittle-to-ductile transition in β- and γ-Al-Mg-phases at 300 °C is known from the literature [26,27]. The microstructure seems to have a significant influence due to the possibility of grain boundary sliding and the effects on the dislocation movement.

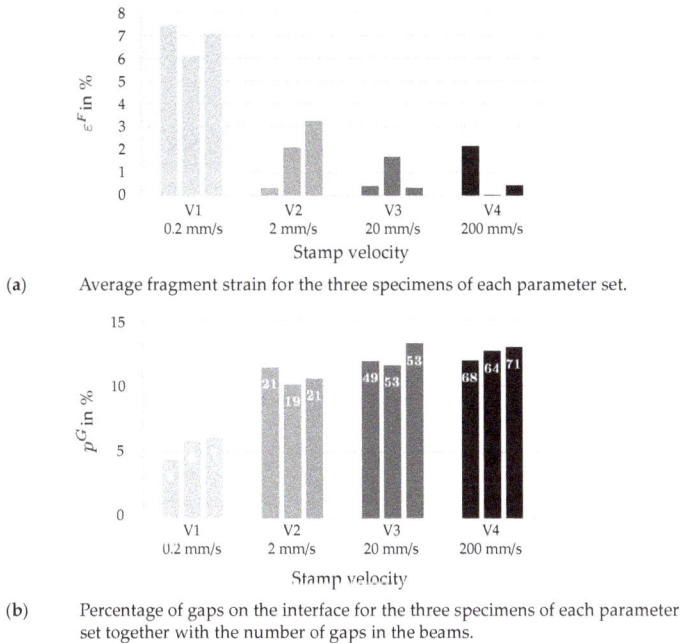

(**a**)   Average fragment strain for the three specimens of each parameter set.

(**b**)   Percentage of gaps on the interface for the three specimens of each parameter set together with the number of gaps in the beams.

Figure 11. Results of the parameter sets with a boundary layer thickness of 25 µm and a varying stamp velocity (**a**) ε^F and (**b**) p^G.

As expected, this trend is also seen by considering the percentage amount of gaps on the interface (Figure 11b). The brittle material behavior expresses itself through an increasing amount of gaps on the interface. The stretch of the interface then arises from the fragment movement and not from the stretch of the fragments.

The comparison of the interface structures in Figure 12 shows two brittle boundary layers. The interface of V4 is represented by a large number of gaps and short fragments, respectively.

The distinctive higher amount of gaps in V4 than in V2 indicates an increasing brittleness with higher stamp velocities even though the difference in the average fragment strain is small.

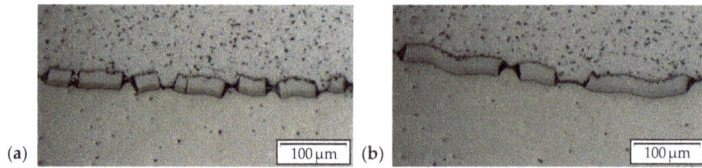

(a) 100 μm (b) 100 μm

Figure 12. Comparison of the fragmentation state in parameter set (**a**) V4 and (**b**) V2.

The dependency of the fragmentation state on the boundary layer thickness is investigated by the same strategy. The associated diagrams are shown in Figure 13. Here again, a clear trend becomes apparent. With a decreasing boundary layer thickness, the ductility increases significantly up to an average fragment strain of 15.6% for a thickness of 7.5 μm.

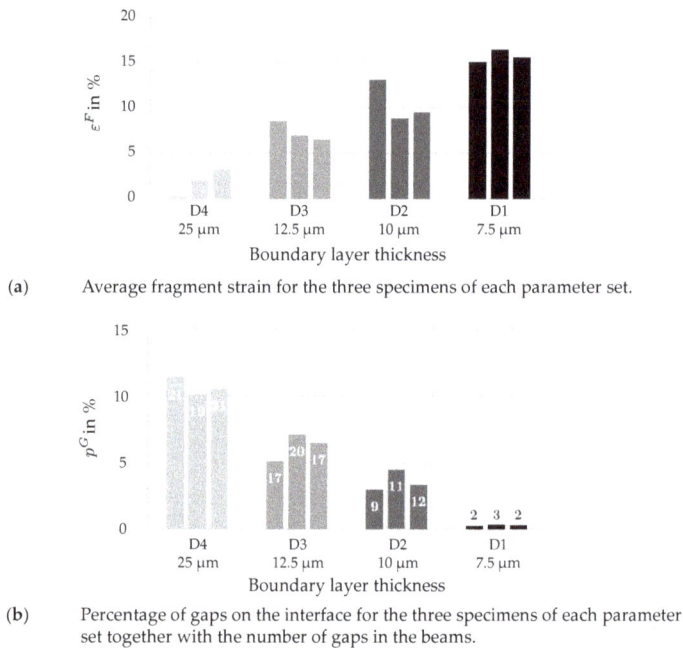

(**a**)   Average fragment strain for the three specimens of each parameter set.

(**b**)   Percentage of gaps on the interface for the three specimens of each parameter set together with the number of gaps in the beams.

Figure 13. Results of the parameter sets with varying boundary layer thickness and a stamp velocity of 2 mm/s (**a**) ε^F and (**b**) p^G.

The percentage of gaps on the interface follows, as expected, a reverse trend compared to the average fragment strain. Decreasing the boundary layer thickness also leads to a decreasing p^G due to an enhanced ductility. The stretch of the interface then results from the stretch of the boundary layer. Firstly remarkable is the extremely low number of gaps for the parameter set D1. Additionally, the gaps seems to be randomly distributed. Secondly, the comparison of the sets D3 (12.5 μm) and D4 (25 μm) reveals an almost identical number of gaps and fragments, respectively. Only the size of the gaps is smaller in D3, therefore indicating a more ductile material behavior. It might be assumed that these gaps emerge at a later point in time in the channel compression test, resulting in the partial filling of the gaps with the basic materials. For a 25 μm thick boundary layer, the interface elongation happens through a movement of the fragments already evolving at the beginning of the compression and resulting in larger gaps.

3.2. Stretch of the Boundary Layer

The computation of the boundary layer stretch is based on the coordinates of the nodes located on the interface between the two basic materials in the numerical simulation. Using the distance of two adjacent nodes δ_i before ($i = 0$) and after ($i = 1$) the compression test respectively, the stretch λ_{BL} between the two nodes can be calculated according to Equation (5):

$$\lambda_{BL} = \frac{\delta_1}{\delta_0} \tag{5}$$

The resulting stretch profile from the simulation is shown in Figure 14 for different loading stages. Due to the specific specimen shape, the deformation starts in the middle section of the interface and propagates further with an increasing stamp shift. At the outer sections, the interface stays in the origin state. This stretch distribution correlates with the observed fragmentation states in the metallographic evaluation (Section 3.1). Furthermore, it can be determined that, within one specimen, stretches from 1 to 1.38 exist at the same time.

As a result, it is not necessary to consider different loading stages to evaluate different elongations of the interface. This circumstance offers the opportunity for obtaining a critical stretch value depending on the fragmentation beginning point as described below in Section 3.3.

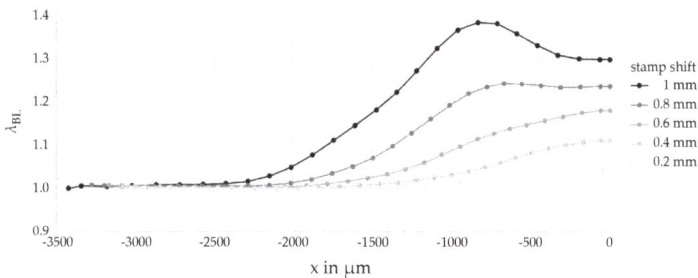

Figure 14. Stretch of the boundary layer λ_{BL} at different loading stages.

It is clear that the stretch maximum is not located in the center of the specimen (stamp shift of 1 mm). In the middle of the specimen, the interface runs horizontal and the channel compression test results in this area in an almost homogenous, simultaneously increasing stretching normal to the compression. The area with the maximum stretch experiences an additional stretching through a rotation of the orientation comparable to a simple shear deformation.

3.3. Fragmentation Criterion

The determination of the critical stretch where the fragmentation of the boundary layer starts is based on the profile variable s. Figure 15 shows the definition of s, which is comparable to the length of a trajectory. It starts at the edge of the specimen and follows the smoothed profile of the interface (dashed line in Figure 15). Smoothing the real profile of the interface improves the comparability to the numerical simulation where the individual meander shape of the real interface is not included.

Figure 15. Definition of the profile coordinates s_L und s_R.

Starting from each specimen edge and following the interface profile, the position of the first gap is marked as $s_{L,crit}$ and $s_{R,crit}$, respectively. At this position, the fragmentation has just started and the stretch at this position can be defined as critical stretch λ_{crit} regarding the fragmentation for the related boundary layer thickness and deformation rate. With the assistance of the numerical simulation, every value of $s_{L,crit}$ and $s_{R,crit}$, respectively, is related to a λ_{crit} value. The results are summarized in Table 2, which contains the average values of λ_{crit} for every parameter set. Additionally, the extreme values $s_{max,crit}$ and $s_{min,crit}$ of $s_{L,crit}$ and $s_{R,crit}$ are listed too for comparison between the parameter sets.

Table 2. Critical stretches of the boundary layer regarding the fragmentation and the measured positions of the fragmentation onset.

Boundary Layer Thickness d	Stamp Velocity	λ_{crit}	$s_{min,crit}$ in μm	$s_{max,crit}$ in μm
25 μm	200 mm/s	1.03	1065	1537
25 μm	20 mm/s	1.04	1228	1615
25 μm	2 mm/s	1.14	1618	2058
25 μm	0.2 mm/s	1.35	2329	3576
7.5 μm	2 mm/s	(1.26)	1524	3979
10 μm	2 mm/s	1.30	2124	2665
12.5 μm	2 mm/s	1.25	1996	2753

In the case of high stamp velocities (200 mm/s and 20 mm/s), the critical stretch amounts to only 1.03 and 1.04, respectively, confirming the statement formulated above that, at high deformation rates and a boundary layer thickness of 25 μm, the elongation at fracture is very small. Decreasing stamp velocity and deformation rate, respectively, lead to an increase of s_{crit} and consequently of the critical stretch. The range with fragmentation in the center of the specimen narrows.

The results of the metallographic evaluation regarding the dependency on the boundary layer thickness are confirmed too with the exception of 7.5 μm. Due to the low quantity of gaps, the determined critical stretch has a poor reliability for this parameter set, for which reason it is excluded from the further utilization. Additionally, the gaps are randomly distributed resulting in inconsistent values of s_{crit}.

On the basis of the experimental and numerical investigations presented above, it is possible to approximate the critical stretch λ_{crit} of the measuring points as a function of the stretch rate $\Delta\lambda / \Delta t$ and the boundary layer thickness d. The mathematical approximation is given by Equation (6). The graphical representation in Figure 16 illustrates the identified dependencies. Distortion conditions under the surface indicate an undamaged boundary layer. Above the surface, fragmentation starts. Therefore, Equation (6) can be interpreted as a fragmentation criterion:

$$\lambda_{crit} = 1.825 - 0.349 \ln d + 0.032 (\ln d)^2 + 0.048\, e^{-\Delta\lambda / \Delta t} + 0.353\, e^{-2\,\Delta\lambda / \Delta t} \tag{6}$$

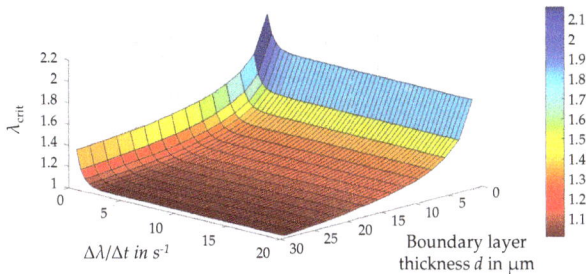

Figure 16. Critical stretch λ_{crit} regarding fragmentation as a function of the stretch rate and the boundary layer thickness.

4. Conclusions

The presented channel compression experiment is very well-suited to investigate the boundary layer deformation behavior in hydrostatic extruded Al-Mg-compounds. Due to the newly developed specimen shape and the resulting inhomogeneous strain distribution, it is possible to detect the onset of fragmentation for a specific deformation rate and boundary layer thickness without a load variation. Thereby, the amount of necessary tests could be reduced considerably. For each parameter set, a critical stretch regarding onset of fragmentation is determined and transferred to a fragmentation criterion. In this context, an initial contribution to the quantification of the load-dependent deformability of the boundary layer is made and further complements the damage model from Feuerhack.

The results indicate the possibility to optimize forging processes regarding damage by fragmentation through a reduction of the process speed and boundary layer thickness. Due to the need for short process times in industry, the reduction of the process speed is limited. On the contrary, the reduction of the boundary layer thickness is possible through shorter heat-up phases during production, making it even more energy efficient respectively.

In future investigations, the three-dimensional deformation of the compound has to be addressed to clarify the mutual influence of two deformation directions. For this purpose, it is recommended to use computer tomography scans for the evaluation.

Acknowledgments: The authors thank the German Research Foundation (DFG) for its funding within the framework of the Collaborative Research Centre 692 (SFB HALS 692).

Author Contributions: Carola Kirbach performed the experimental and numerical investigations and wrote the paper. Martin Stockmann, Jörn Ihlemann and Carola Kirbach discussed the results.

Conflicts of Interest: The authors declare no conflict of interest.

References

1. Klüting, M.; Landerl, C. The new BMW inline six-cylinder spark-ignition engine. *MTZ Worldw.* **2004**, *65*, 2–5.
2. Kleiner, M.; Schomäcker, M.; Schikorra, M.; Klaus, A. Herstellung verbundverstärkter Aluminiumprofile für ultraleichte Tragwerke durch Strangpressen. *Materialwissenschaft und Werkstofftechnik* **2004**, *35*, 431–439.
3. Avitzur, B.; Wu, R.; Talbert, S.; Chou, Y. An Analytical Approach to the Problem of Core Fracture During Extrusion of Bimetal Rods. *ASME. J. Eng. Ind.* **1985**, *107*, 247–253.
4. Story, J.; Avitzur, B.; Hahn, W.C., Jr. The Effect of Receiver Pressure on the Observed Flow Pattern in the Hydrostatic Extrusion of Bimetal Rods. *ASME. J. Eng. Ind.* **1976**, *98*, 909–913.
5. Feuerhack, A. Experimentelle und Numerische Untersuchungen von Al-Mg-Verbunden Mittels Verbundschmieden. Ph.D. Thesis, Chemnitz University of Technology, Chemnitz, Germany, 23 May 2014.
6. Lehmann, T. Experimentell-Numerische Analyse Mechanischer Eigenschaften von Aluminium/Magnesium-Werkstoffverbunden. Ph.D. Thesis, Chemnitz University of Technology, Chemnitz, Germany, 29 June 2012.
7. Kittner, K.; Feuerhack, A.; Förster, W.; Binotsch, C.; Graf, M. Recent Developments for the Production of Al-Mg Compounds. *Mater. Today Proc.* **2015**, *2*, S225–S232.

8.   Kirbach, C.; Stockmann, M.; Ihlemann, J. Experimental and numerical investigations of Al/Mg-compound specimens under bending after mass forming processes. In Proceedings of the 31st Danubia Adria Symposium on Advances in Experimental Mechanics, Kempten, Germany, 24–27 September 2014; pp. 32–33.
9.   Förster, W.; Binotsch, C.; Awiszus, B.; Lehmann, T.; Müller, J.; Kirbach, C.; Stockmann, M.; Ihlemann, J. Forging of eccentric co-extruded Al-Mg compounds and analysis of the interface strength. *IOP Conf. Ser. Mat. Sci. Eng.* **2016**, *118*, 012032.
10. Kosch, K.G.; Frischkorn, C.; Huskic, A.; Odening, D.; Pfeiffer, I.; Prüß, T.; Vahed, N. *Effizienter Leichtbau Durch Belastungsangepasste und Anwendungsoptimierte Multimaterial-Schmiedebauteile*; UTF Science, Meisenbach Verlag Bamberg: Bamberg, Germany, 2012; pp. 1–17.
11. Kosch, K.G.; Pfeiffer, I.; Foydl, A.; Behrens, B.; Tekkaya, A. *Schmieden von Partiell Stahlverstärkten Aluminiumhalbzeugen*; UTF Science, Meisenbach Verlag Bamberg: Bamberg, Germany, 2012; pp. 1–9.
12. Foydl, A.; Pfeiffer, I.; Kammler, M.; Pietzka, D.; Matthias, T.; Jäger, A.; Tekkaya, A.; Behrens, B. Manufacturing of Steel-Reinforced Aluminium Productes by Combining Hot Extrusion and Closed-Die Forging. *Key Eng. Mater.* **2012**, *504–506*, 481–486.
13. Kittner, K. Integrativer Modellansatz bei der Co-Extrusion von Aluminium-Magnesium-Werkstoffverbunden. Ph.D. Thesis, Chemnitz University of Technology, Chemnitz, Germany, 11 May 2012.
14. Negendank, M.; Mueller, S.; Reimers, W. Coextrusion of Mg–Al macro composites. *J. Mater. Process. Technol.* **2012**, *212*, 1954–1962.
15. Lehmann, T.; Stockmann, M. Residual Stress State and Fracture Mechanical Properties of Al/Mg Compounds. *Mater. Today Proc.* **2016**, *3*, 1041–1044.
16. Lehmann, T.; Stockmann, M. Residual Stress Analysis of Al/Mg Compounds by Using the Hole Drilling Method. *J. Jpn. Soc. Exp. Mech.* **2011**, *11*, s233–s238.
17. Lehmann, T.; Stockmann, M.; Naumann, J. Experimental and Numerical Investigations of Al/Mg Compound Specimens under Load in an Extended Temperature Range. *FME Trans.* **2009**, *37*, 1–8.
18. Kittner, K.; Awiszus, B.; Lehmann, T.; Stockmann, M.; Naumann, J. Numerische und experimentelle Untersuchungen zur Herstellung von stranggepressten Aluminium/Magnesium-Werkstoffverbunden und zur Festigkeit des Interface. *Materialwissenschaft und Werkstofftechnik* **2009**, *40*, 532–539.
19. Kittner, K.; Binotsch, C.; Awiszus, B.; Lehmann, T.; Stockmann, M. Herstellungsprozess zur Erzeugung schädigungsarmer Al/Mg-Verbunde und Analyse der mechanischen Grundeigenschaften sowie der Interfacefestigkeit. *Materialwissenschaft und Werkstofftechnik* **2010**, *41*, 744–755.
20. Lehmann, T.; Stockmann, M.; Kittner, K.; Binotsch, C.; Awiszus, B. Bruchmechanische Eigenschaften von Al/Mg-Verbunden und deren Fließverhalten im Herstellungsprozess. *Materialwissenschaft und Werkstofftechnik* **2011**, *42*, 612–623.
21. Kirbach, C.; Lehmann, T.; Stockmann, M.; Ihlemann, J. Digital Image Correlation Used for Experimental Investigations of Al/Mg Compounds. *Strain* **2015**, *51*, 223–234.
22. Dietrich, D.; Nickel, D.; Krause, M.; Lampke, T.; Coleman, M.P.; Randle, V. Formation of intermetallic phases in diffusion-welded joints of aluminium and magnesium alloys. *J. Mater. Sci.* **2011**, *46*, 357–364.
23. Pawelski, H. Erklärung Einiger Mechanischer Eigenschaften von Elastomerwerkstoffen mit Methoden der Statistischen Physik. Ph.D. Thesis, Universität Hannover, Hannover, Germany, 1998.
24. Feuerhack, A.; Binotsch, C.; Awiszus, B.; Wolff, A.; Brämer, C.; Stockmann, M. Materialeigenschaften und Formänderungsvermögen von stranggepressten Al-Mg-Verbunden in Abhängigkeit der Temperatur. *Materialwissenschaft und Werkstofftechnik* **2012**, *43*, 601–608.
25. Noster, U.; Scholtes, B. Isothermal strain-controlled quasi-static and cyclic deformation behavior of magnesium wrought alloy AZ31. *Zeitschrift für Metallkunde* **2003**, *94*, 559–563.
26. Ragani, J.; Donnadieu, P.; Tassin, C.; Blandin, J. High-temperature deformation of the γ-Mg17Al12 complex metallic alloy. *Scr. Mater.* **2011**, *65*, 253–256.
27. Roitsch, S. Microstructural and macroscopic aspects of the plasticity of complex metallic alloys. Ph.D. Thesis, RWTH Aachen University, Aachen, Germany, 2008.

![metals logo] *metals*

MDPI

Article

Arc Brazing of Aluminium, Aluminium Matrix Composites and Stainless Steel in Dissimilar Joints

Thomas Grund [1,*], Andreas Gester [1], Guntram Wagner [1], Stefan Habisch [2] and Peter Mayr [2]

[1]   Institute of Materials Science and Engineering, Chemnitz University of Technology, 09127 Chemnitz, Germany; andreas.gester@mb.tu-chemnitz.de (A.G.); guntram.wagner@mb.tu-chemnitz.de (G.W.)
[2]   Institute of Joining and Assembly, Chemnitz University of Technology, 09126 Chemnitz, Germany; stefan.habisch@mb.tu-chemnitz.de (S.H.); peter.mayr@mb.tu-chemnitz.de (P.M.)
*   Correspondence: thomas.grund@mb.tu-chemnitz.de; Tel.: +49-371-531-35390

Received: 30 January 2018; Accepted: 5 March 2018; Published: 8 March 2018

Abstract: The publication describes the approaches and results of the investigation of arc brazing processes to produce dissimilar joints of particle reinforced aluminium matrix composites (AMC) to aluminium alloys and steels. Arc brazing allows for low thermal energy input to the joint parts, and is hence suitable to be applied to AMC. In addition, a braze filler B-Al40Ag40Cu20 alloyed with Si with a liquidus temperature of below 500 °C is selected to further reduce the thermal energy input during joining. The microstructures of the joining zones were analysed by scanning electron microscopy (SEM), energy dispersive X-ray spectroscopy (EDXS), and X-ray diffraction analysis (XRD), as well as their hardness profile characterised and discussed. Joint strengths were measured by tensile shear tests, and resulting areas of fracture were discussed in accordance to the joints' microstructures and gained bond strength values.

Keywords: arc brazing; brazing fillers; microstructure; wettability; aluminium matrix composite; AMC; stainless steel; aluminium; joining

---

1. Introduction

Dissimilar metal joints under presence of light metals show a high potential to further improve existing light weight solutions, since they strongly widen the design flexibility of components and assemblies. To this design flexibility, high performance materials, like aluminium matrix composites (AMC), add further advantages regarding the mechanical and physical properties, like a lower coefficient of thermal expansion, higher specific strength, or an increased wear resistance in comparison to non-reinforced aluminium [1–3]. The challenges of thermally joining aluminium to steel are known. A large difference between their melting points, poor solid solubility of iron in aluminium, and hence, the formation of Al-Fe intermetallic phases lower the resulting bond strengths of such joints [4–7]. Further bond strength reduction results from stress induced cracks alongside the formed brittle phases [8,9]. Hence, different thermal joining techniques with lowered heat input to the joining area or reduced joining times are a permanent object of research and development. Thus, friction welding, spot welding, laser-assisted welding, and brazing processes, as well as hybrid welding processes are suitable techniques to produce resilient dissimilar Al/steel joints [10–13]. The herein focussed arc brazing represents a cost-efficient alternative to laser-assisted welding and brazing. When AMC partners are subjected to thermal joining processes, their low thermal stability must additionally be considered. The matter becomes more pronounced, when the reinforcing particle phase comprise SiC that either thermally degrades or dissolves into the Al matrix, if the thermal load from the joining process is too high. Dissolving will result in brittle intermetallic Al_3C_4 precipitates and a joint strength decrease of 50% [14,15]. Furthermore, the level and time of the heat input influence metallurgic processes at the interface between matrix and reinforcement phase and can lead to high porosity levels and chemical inhomogeneity in joining zones, if not adapted [16].

The presented work faces the above described problems. The work aim is an improved structural homogeneity of joint zones and hence higher joint strengths of dissimilar AMC/Al and AMC/steel joints in comparison to those resulting from conventional thermal joining techniques, like fusion welding, flux-aided furnace, or vacuum brazing. To achieve this aim, arc brazing is applied. Arc brazing offers the possibility to effectively minimize the heat input to joining zones. Especially, the thermal damage of AMC microstructures shall to be avoided by the combination of a focused arc with localised heat input and a low-melting braze filler. Another benefit is expected from an improved wetting of the braze filler on the substrates due to a process-inherent degradation of oxide layers at the surfaces of the metallic joining partners. The described aim could also be achieved by non-fusion welding, like friction welding, but arc brazing is chosen for its higher design flexibility regarding joint zone geometries and the non-existent wear of joining tools when processing AMC with hard-phase reinforcement.

2. Materials and Methods

The used AMC joining partners are from powder-metallurgically produced material EN AW-2017 + 10 vol % (SiC)$_P$. Stainless steel joining partners are sheets from AISI 304L, while aluminium partners are made from EN AW-6082. All surfaces to be joined are ground and polished to $R_z < 1$ µm. Arc brazing processes are carried out using a TIG (tungsten inert gas) welding unit with alternating current (AC) of 40 A and Ar shielding gas (EWM, Muendersbach, Germany). Test specimens are brazed as lap joints. The braze filler is applied manually. In wetting tests, the arc is ignited next to an applied braze filler pearl of 0.1 g mass and subsequently moved over it. The arc is interrupted either after wetting occurred or after 4 s of process time to avoid melting of the base material. During joining and wetting experiments, the work distance between electrode and sample surface is approximately 10 mm. Microstructure investigations are performed by light microscopy (LM, Olympus Europa, Hamburg, Germany), scanning electron microscopy (SEM, LEO1455VP, Carl Zeiss Microscopy, Jena, Germany), and hardness measurements according to Vickers with a force of 0.005 kp (i.e., a load of about 5 g) with an imprint duration of 15 s (HV0.005/15). The chemical composition of phases and larger areas like diffusion zones are examined via energy dispersive X-ray spectroscopy (EDXS, EDAX Genesis, Carl Zeiss Microscopy, Jena, Germany) and X-ray diffraction analysis (XRD, D8 "Discover", Bruker, Billerica, MA, USA) with Co-Kα radiation (wavelength $\lambda = 0.178886$ nm, phase database PDF-2014). Wetting angles are measured at cross sections images, using a digital imaging tool. Differential scanning calorimetry (DSC, Netzsch, Selb, Germany) is used to determine the melting temperature of applied braze fillers. Produced lap joints are measured regarding shear tensile strengths at test speed $v = 1$ mm/min and room temperature.

The commonly used braze fillers for brazing aluminium alloys are Al-Si- or Zn-Al-based alloys [4,17–20]. Beside a better corrosion resistance of the Al-Si braze fillers, in dissimilar Al/steel joints Si prevents the growth of intermetallic Fe-Al phases by lowering the diffusivity of aluminium due to the formation of intermetallic Fe-Al-Si phases [21]. On the downside, the working temperature of Al-Si braze fillers is higher in comparison to Zn-Al fillers and may exceed the solidus temperature of Al-Si-based aluminium casting alloys that are frequently used in AMC. Therefore, a recently introduced braze filler basing on a ternary Al-Ag-Cu system comprising the eutectic composition of 40 wt % Al, 40 wt % Ag and 20 wt % Cu is used within the presented work (B-Al40Ag40Cu20, $T_m = 506$ °C [22]). This braze filler was alloyed with Si to improve its wetting behaviour on stainless steel [23]. Therefore, a master alloy (B-Ag72Cu-780) is continuously casted and complemented with Al (purity 99.99%) and Cu (purity 99.9%) to gain the announced eutectic composition. Furthermore, 1.2–1.4 wt % of Si are added (purity 99.9999%). To prevent melt pool oxidation, the braze filler production is done in inert Ar atmosphere. Besides positively affecting the braze filler's wetting behaviour on stainless steel, adding Si reduces its melting temperature from 506 °C to 497 °C in comparison to the former ternary eutectic composition, Table 1. The homogeneity of alloying components within the produced braze filler rods is controlled and proven by EDXS analysis.

Table 1. Chemical compositions and melting temperatures of the eutectic and Si-alloyed braze fillers.

Specification	Chemical Composition (wt %)				T_M (°C)
	Al	Ag	Cu	Si	
Eutectic Al-Ag-Cu braze filler (B-Al40Ag40Cu20)	40	40	20	-	506
Si-alloyed Al-Ag-Cu braze filler (B-Al39.5Ag39.5Cu20Si1)	39.44–39.52	39.44–39.52	19.72–19.76	1.2–1.4	497

3. Results and Discussion

3.1. Wetting Behaviour

The boarders between different categories of wetting behaviour are openly discussed and often related to a specific field of application. Distinctions between "good wetting" and "poor wetting", as well as "wetting" and "no wetting", however, are frequently fixed to wetting angles of either less or more than 90° [24–27]. For brazing processes, an optimal wetting angle between liquid braze filler and solid joining partner is often defined as less than 30° [24,27,28], but sufficient diffusion between braze filler and base material takes already place at much higher wetting angles. B-Al40Ag40Cu20 braze filler shows very good wetting when arc brazed on AMC, with wetting angles of less than 20°. Still satisfactory wetting is documented for it on stainless steel, with wetting angles of approximately 90°. As presented in a former publication, a brazing current variation between 40 A and 45 A does not visibly influence the wetting result [29]. SEM/EDXS analyses of AMC wetting samples show large diffusion zones. Qualitative XRD suggests a dominant ternary eutectic microstructure that results from the braze filler and consists of $CuAl_2$, Ag_2Al, as well as Al solid solution. Close to the AMC interface, primary solidified Al- and Al-Ag-rich solid solutions are identified, Figures 1 and 2. XRD further reveals a content of $(SiC)_P$ reinforcement that dissolved from the AMC base material as well as intermetallic Al_7Ag_3. Details are presented in and Table 2 and in [29]. Since EDXS measurements include data from a globular excitation volume of about 2 μm in diameter, the presented chemical compositions of the analysed fine phases represent qualitative results, and were hence rounded to integral numbers. The results are backed up by literature data [30,31].

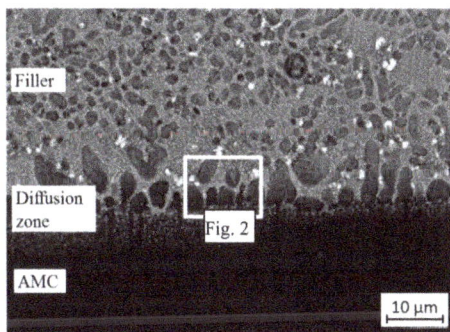

Figure 1. Microstructure (SEM) of braze filler B-Al40Ag40Cu20 wetted on AMC EN AW-2017 + 10 vol % (SiC)P, current intensity 40 A [29].

Figure 2. Braze seam of area marked in Figure 1, 1: ternary eutectic microstructure, 2: primary solidified Al-rich solid solution, 3: primary solidified Al-Ag-rich solid solution, 4: ternary eutectic microstructure [29].

Table 2. Chemical compositions (energy dispersive X-ray spectroscopy (EDXS)) of phases and detected phases (X-ray diffraction analysis (XRD)) in braze filler B-Al40Ag40Cu20 wetted on aluminium matrix composites (AMC) EN AW-2017 + 10 vol % (SiC)$_P$ in accordance to markings in Figure 2.

Phase No. in Accordance to Figure 2	Chemical Composition by EDXS (at %)			
	Al	Ag	Cu	Si
1	84	10	6	-
2	80	14	6	-
3	62	30	8	-
4	81	9	6	4

Phases (qualitative XRD):

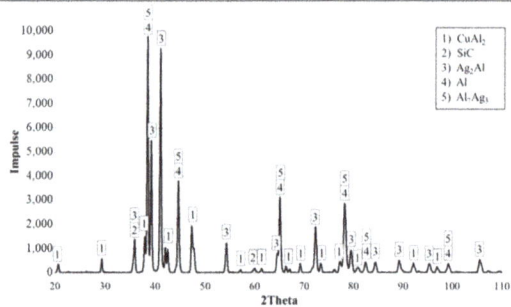

In comparison to these results, stainless steel samples wetted by B-Al40Ag40Cu20 show a compact intermetallic phase seam next to the steel interface with columnar solidified crystal structures growing in the direction of the eutectic microstructure of the braze filler, Figures 3 and 4. Compact hard phase seams can lower the resulting bond strength significantly [32]. In the case of the present material combination, thickly developed brittle seams are prone to stress cracking during joint cooling, which is induced by different coefficients of thermal expansion (CTE).

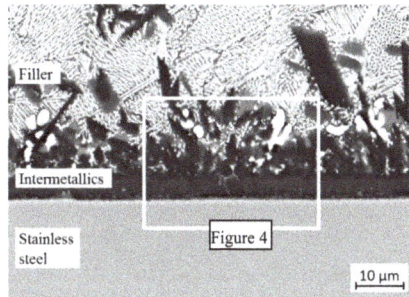

Figure 3. Microstructure (SEM) of braze filler B-Al40Ag40Cu20 wetted on stainless steel AISI 304L, braze current intensity 40 A [29].

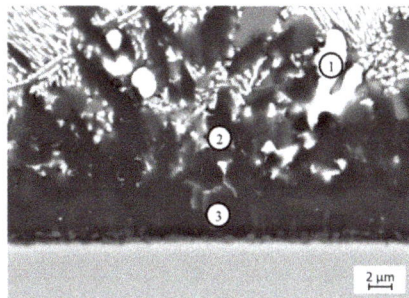

Figure 4. Braze seam of area marked in Figure 3, 1: Al-Ag-rich solid solution, 2: Fe-containing Al-rich phase (probably Al13Fe4), 3: intermetallic phase seam of Al-Fe type with contents of Cr and Ni [29].

In the present material combination, the CTE values are 23.4×10^{-6} K^{-1} and 18×10^{-6} K^{-1} for the braze filler and base material, respectively [33]. EDXS and XRD analyses, again being presented in detail in [29], define the columnar structures as large areas of an Al-Ag-rich phase next to a Fe-containing Al-rich phase, whose chemical composition indicates intermetallic Al$_{13}$Fe$_4$. The compact phase seam probably consists of the intermetallic Al$_2$Fe containing Cr and Ni from the base material. In addition a solid solution of Fe, Ni, and Cu with austenitic lattice is determined near the steel interface [29,34]. Details are presented in and Table 3. Again, the therein given EDXS values were rounded to integral numbers.

3.2. Dissimilar Joints by Arc Brazing

With regard to the Al-based joining partners, the interfaces between the processed B-Al39.5Ag39.5Cu20Si1 braze filler and the respective base material that develop during the applied arc brazing processes are similar to those that are resulting from the wetting tests. According to SEM investigations, AMC and non-reinforced aluminium EN AW-6082 both exhibit a eutectic braze seam microstructure resulting from the braze filler. Hence, the fine eutectic microstructure is interpreted as comprising the ternary eutectic phases Al$_2$Cu, Ag$_2$Al, as well as Al solid solution. At the EN AW-6082 surface an additional seam of Al solid solution is detected. This non-cohesive phase shows a higher amount of Al compared to the Al solid solution of the ternary eutectic microstructure that resulted from the wetting test, Figure 5a. In comparison to this, the braze seam microstructure close to the AMC also shows a non-cohesive Al phase seam, though with lower homogeneity. In addition, due to the composition of the AMC EN AW-2017 matrix, which is closer to the composition of the braze filler in comparison to aluminium EN AW-6082, the diffusivity and formation of the ternary eutectic phases Al$_2$Cu, Ag$_2$Al, and Al solid solution is slightly increased. This results in a more pronounced diffusion zone of approximately twice the thickness of that on EN AW-6082, Figure 5b. The additionally observed Cu precipitates in the AMC base material are interpreted as a result of the heat input during brazing and a subsequent precipitation hardening effect.

Table 3. Chemical compositions (EDXS) of phases and detected phases (XRD) in braze filler B-Al40Ag40Cu20 wetted on stainless steel AISI 304L in accordance to markings in Figure 4.

Phase No. in Accordance to Figure 4	Chemical Composition by EDXS (at %)					
	Al	**Ag**	**Cu**	**Fe**	**Cr**	**Ni**
1	52	40	5	2	1	-
2	75	1	3	17	3	1
3	52	-	2	33	9	4

Phases (qualitative XRD):

| (a) | (b) | (c) |

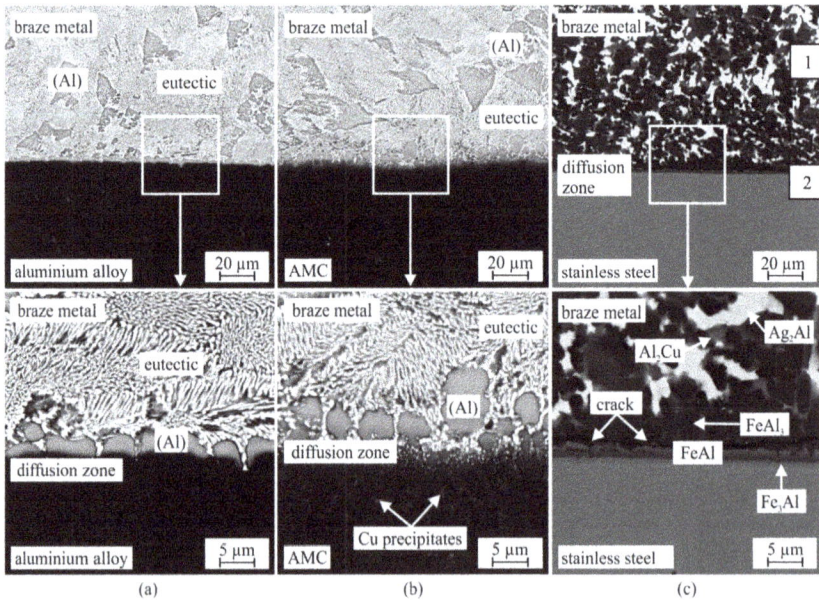

Figure 5. Microstructure (SEM) of the interfaces of braze seams between B-Al39.5Ag39.5Cu20Si1 braze filler and (**a**) aluminium EN AW-6082, (**b**) AMC EN AW-2017 + 10 vol % (SiC)$_P$ as well as (**c**) stainless steel AISI 304L [35].

The braze seam microstructure of the interface close to stainless steel exhibits obvious differences when compared to those of Al-based joint partners as well as the results from wetting tests, Figure 5c. The respective phase detection by XRD measurements is presented in Table 4 in accordance to the marked regions in Figure 5c (above). The ternary eutectic microstructure of the braze filler is not existent, although Al_2Cu and Ag_2Al are found by qualitative XRD measurements in this region. The diffusion zone between braze seam and base material comprises at least three consistent phase layers, forming a seam of about 10 μm total thickness. Qualitative XRD measurements in the interface area revealed three intermetallic phases Fe_3Al, FeAl, and $Al_{37}Cu_2Fe_{12}$ (Al_3Fe-type), which are interpreted to form this visible seam with an increasing share of Al in the direction of the braze seam. Detailed analysis results are presented by the authors in [35]. In SEM images, stress induced cracks are visible in the interface between the FeAl and $FeAl_3$ layer, Figure 5c below. This failure already occurs in braze seams not subjected to mechanical load. Hence, the crack initiation can be lead back to abrupt changes in the hardness levels (FeAl: 470 HV1, $FeAl_3$: 892 HV1) and combined internal stresses during cooling [36]. The detection of $FeAl_3$ in the braze seam far from the interface to the steel base material (comp. Table 4) indicates a strong alloying of the braze seam with Fe, probably due to erosion of the steel base material.

Table 4. Detected phases (XRD) in the interface of braze seams between B-Al39.5Ag39.5Cu20Si1 braze filler and stainless steel AISI 304L in accordance to region markings in Figure 5c (above).

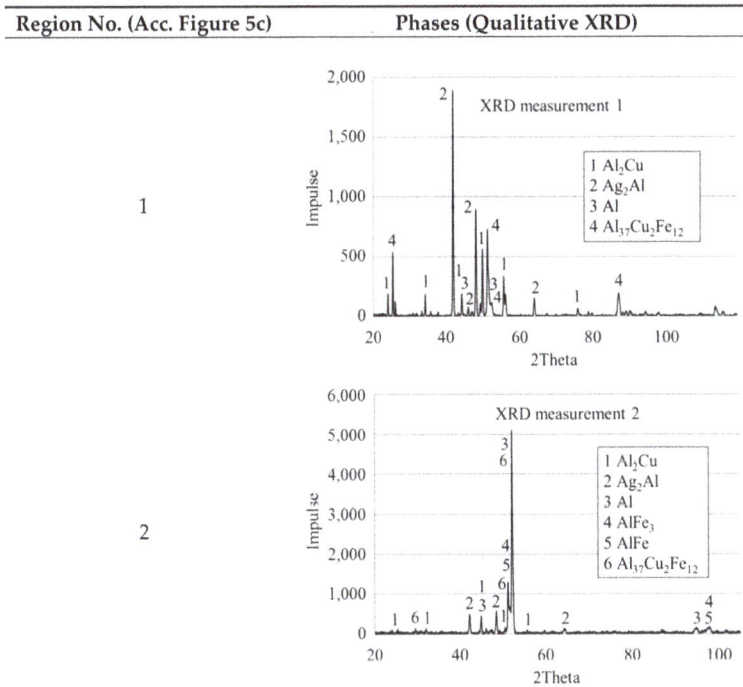

Region No. (Acc. Figure 5c)	Phases (Qualitative XRD)
1	
2	

The different microstructure features of the interfaces between B-Al39.5Ag39.5Cu20Si1 braze seam and the different base materials are also visible in their respective hardness profiles, Figure 6. The displayed profiles were taken within a distance of 30 μm from the braze seam/base material interface in either direction. Hardness levels are recorded in steps of 10 μm. To ensure the demanded distance between the measuring points according to Vickers HV0.005/15, measurements are taken along an offset line (cross section picture in Figure 6). A smooth hardness transition is documented between braze seam and Al-based joining partners, resulting from the Al solid solution phase that forms close to the respective interfaces. These nearly constant hardness profiles are considered to significantly lower the risk of cracks in this area, when mechanical load is applied. In contrast to this observation, the hardness increase at the interface of braze seam and stainless steel is much stronger. This results from the above documented alloying of the braze seam with Fe, Ni, and Cr from the steel base material and the subsequent formation of intermetallic Al-Fe phases. In the direction of the former steel surface, the content of Al, and hence the measured hardness levels decrease from about 1000 HV0.005/15 to 500 HV0.005/15. In addition, the non-eutectic microstructure of the braze seam in this area exhibits a significantly higher hardness level than that close to the Al base materials.

The features and properties of the different microstructures in the investigated dissimilar joints become obvious in the conducted tensile shear tests. Tensile shear test samples were produced from sample sheets with height $H = 1.5$ mm, width $W = 10$ mm, and lengths of $L_1 \sim 50$ mm. The joint area overlap was $L_2 = 15$ mm. Due to the arc brazing joint geometry, however, true joining areas differed from each other according to applied braze filler amount and the wettability of the base materials. The respective joining area ranges for the different dissimilar joints are given in Table 5.

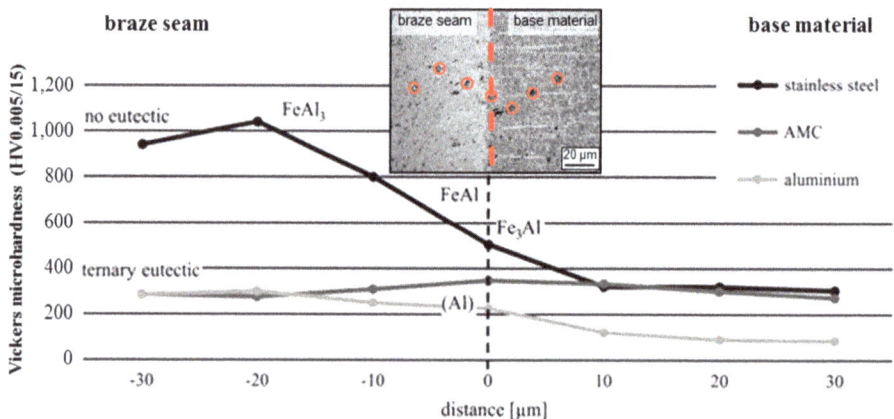

Figure 6. Hardness profiles at the investigated interfaces in dissimilar joints [35].

Figure 7 shows tensile shear strength values obtained from respective tests of braze joints of AMC EN AW-2017 + 10 vol % (SiC)$_P$ to itself and in dissimilar joints to EN AW-6082 and steel AISI304L. Figure 8 shows the tested specimens and areas of fracture. The developed arc brazing routine under use of the newly introduced braze filler B-Al39.5Ag39.5Cu20Si1 permits AMC/AMC joints with average tensile shear strengths of about 55 MPa, showing maximum values of about 80 MPa. The resulting areas of fracture comprise partly ductile failure that mainly occur within the braze seam material close to the AMC joining partner with lower braze filler wetted area. The ductile failure is attributed to the well-developed diffusion zone between the braze seam and AMC base material in this region. In comparison to flux-aided furnace brazing processes the bond strength values are high. This is attributed to the successful oppression of (SiC$_P$) dissolution and the related formation of aluminium carbides. Applying the same arc brazing routine and braze filler to dissimilar AMC/EN AW-6082 joints results in lower average bond strengths of about 40 MPa. Fracture occurs close to the surface of the EN AW-6082 base material. This is lead back to the underrepresented diffusion zones on the former Al alloy surfaces. The pictured fracture area in Figure 7 hence shows less ductile zones in comparison to the fracture area of AMC/AMC samples. For the applied arc brazing routine, the bond strength of dissimilarly joined AMC/AISI304L is lowest in comparison to the tensile shear strengths of

AMC/AMC and AMC/EN AW-6082 brazing joints. The average value is about 20 MPa, which leads back to the microstructure, hardness level, and crack density that occur in the braze seam area close to the steel partner. Hence, brittle fracture occurs at the surface of the AISI304L base material. The same place of failure occurs in flux-aided brazing joints of the same type. However, in comparison to results from conventional flux-aided furnace brazing, the average bond strength value is doubled. This positive result is understood to come from a comparatively thinner seam of intermetallic phases and a finer dispersed distribution of intermetallic phases within the braze seam close to the former steel surface, when the described arc brazing routine is applied.

Table 5. True joining areas of dissimilar joints tested regarding shear tensile strength (Figures 7 and 8).

Braze Joint Type	AMC/AMC	AMC/EN-6082	AMC/AISI304L (Arc Brazed)	AMC/AISI304L (Flux Brazed)
Joining Area (mm^2)	33–36.5	26–51	34–112	49–85

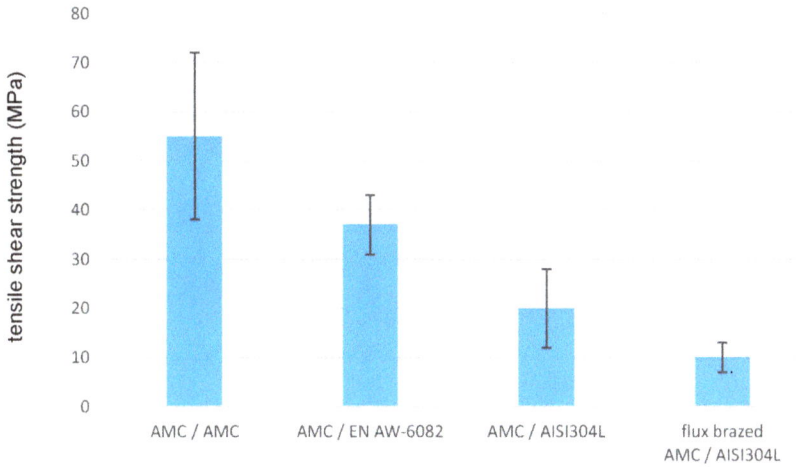

Figure 7. Bond strength values resulting from tensile shear tests for different similar and dissimilar braze joints using arc brazing routine and braze filler B-Al39.5Ag39.5Cu20Si1.

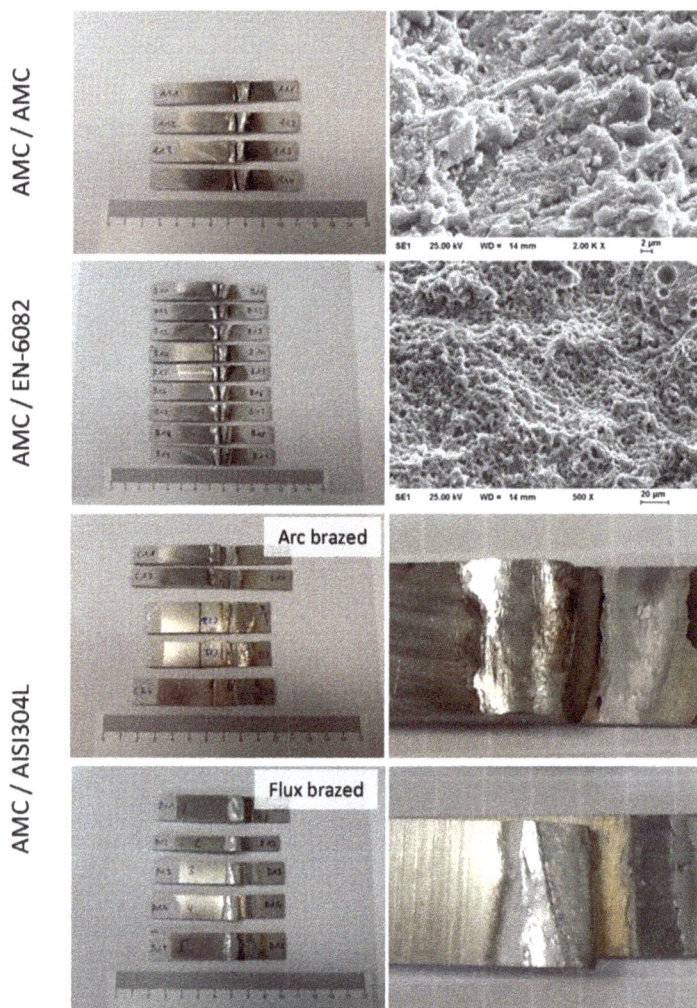

Figure 8. Tested specimens and resulting areas of fracture in accordance to the test and bond strength values presented in Figure 7.

4. Conclusions

Joints of particle reinforced aluminium matrix composites to itself, stainless steel, and aluminium alloy were successfully produced by arc brazing. The investigation of the microstructures of the developed braze filler and the produced joints proved that the resulting thermal influence on the processed aluminium materials is low, which results in thin diffusion zones. Additionally, it is noticeable that the ternary eutectic microstructure of the braze filler still exists within the braze seam close to the aluminium-based joining partners. The hardness profile at the interfaces between braze seam and aluminium alloy as well as AMC is accordingly flat, showing nearly constant hardness levels of approximately 300 HV0.005/15. The nearly constant hardness levels lead to high average bond strengths of these brazing joints of about 40 MPa and 55 MPa, respectively. In contrast to this, the interfaces of the braze seam to stainless steel showed detrimental microstructures and hardness profiles, due to severe alloying of the braze filler during arc brazing. As a result, various intermetallic phases and phase seams formed instead of the announced ternary eutectic microstructure that was documented for braze seam areas that are close to the Al-based joining partners. However, the resulting average bond strength of 20 MPa was twice as high as that gained for comparatively produced conventionally furnace-brazed AMC/steel joints.

In summary, arc brazing is a suitable technique to join AMC to Al alloys with high bond strengths. Permanent arc brazed joints of AMC to steel could be produced, however, shear tensile tests revealed significantly lower bond strengths due to a persistent brittle phase seam at the surface of the steel partner. Hence, further adapted brazing processes are necessary to reduce heat input and to prevent the formation of brittle intermetallic phases, especially seams of intermetallic Fe-Al phases.

Acknowledgments: The authors gratefully acknowledge the funding by the German Research Foundation (Deutsche Forschungsgemeinschaft, DFG) within the framework of the Collaborative Research Centre 692 (SFB HALS 692).

Author Contributions: Thomas Grund and Andreas Gester finished the joining experiments and the investigations and characterization of the joint properties. Thomas Grund is also the main author of this manuscript. Thomas Grund and Peter Mayr are senior scientists, who developed and supervised the project "Joining concepts for bulk and sheet metal structures of high-strength light-weight materials" within SFB HALS 692. Guntram Wagner is a senior scientist who holds the chair of the Composite Materials and Material Compounds group and advised and supervised the brazing experiments.

Conflicts of Interest: The authors declare no conflicts of interest.

References

1. Xiu, Z.; Yang, W.; Chen, G.; Jiang, L.; Mac, K.; Wu, G. Microstructure and tensile properties of Si_3N_{4p}/2024Al composite fabricated by pressure infiltration method. *Mater. Des.* **2012**, *33*, 350–358. [CrossRef]
2. Sajjadi, S.A.; Ezatpour, H.R.; Parizi, M.T. Comparison of microstructure and mechanical properties of A356 aluminium alloy/Al_2O_3 composites fabricated by stir and compo-casting processes. *Mater. Des.* **2012**, *34*, 106–111. [CrossRef]
3. Qu, X-h; Zhang, L.; Wu, M.; Ren, S.-b. Review of metal matrix composites with high thermal conductivity for thermal management applications. *Prog. Nat. Sci. Mater. Int.* **2011**, *21*, 189–197.
4. Song, J.L.; Lin, S.B.; Yang, C.L.; Ma, G.C.; Liu, H. Spreading behavior and microstructure characteristics of dissimilar metals TIG welding-brazing of aluminum alloy to stainless steel. *Mater. Sci. Eng. A* **2009**, *509*, 31–40. [CrossRef]
5. Uzun, H.; Dalle Donne, C.; Argagnotto, A.; Ghidini, T.; Gambaro, C. Friction stir welding of dissimilar Al 6013-T4 to X5CrNi18-10 stainless steel. *Mater. Des.* **2005**, *26*, 41–46. [CrossRef]
6. Liu, H.W.; Guo, C.; Cheng, Y.; Liu, X.F.; Shao, G.J. Interfacial strength and structure of stainless steel–semi-solid aluminum alloy clad metal. *Mater. Lett.* **2006**, *60*, 180–184. [CrossRef]
7. Steiners, M.; Höcker, F. Einfluss der Beschichtungen beim stoffschlüssigen Lichtbogen-fugen von Stahl mit Aluminium. *Mater. Werkst.* **2007**, *38*, 559–564. [CrossRef]
8. Dong, H.; Yang, L.; Dong, C.; Kou, S. Arc joining of aluminum alloy to stainless steel with flux-cored Zn-based filler metal. *Mater. Sci. Eng. A* **2010**, *527*, 7151–7154. [CrossRef]
9. Lin, S.B.; Song, J.L.; Yang, C.L.; Fan, C.L.; Zhang, D.W. Brazability of dissimilar metals tungsten inert gas butt welding-brazing between aluminum alloy and stainless steel with Al–Cu filler metal. *Mater. Des.* **2010**, *31*, 2637–2642. [CrossRef]
10. Ding, Y.; Shen, Z.; Gerlich, A.P. Refill friction stir spot welding of dissimilar aluminum alloy and AlSi coated steel. *J. Manuf. Proc.* **2017**, *30*, 353–360. [CrossRef]
11. Sakiyama, T.; Murayama, G.; Naito, Y.; Saita, K.; Miyazaki, Y.; Oikawa, H.; Nose, T. Dissimilar Metal Joining Technologies for Steel Sheet and Aluminum Alloy Sheet in Auto Body. *Nippon Steel Technol. Rep.* **2013**, *103*, 91–98.
12. Casalino, G.; Leo, P.; Mortello, M.; Perulli, P.; Varone, A. Effects of Laser Offset and Hybrid Welding on Microstructure and IMC in Fe-Al Dissimilar Welding. *Metals* **2017**, *7*, 282. [CrossRef]
13. Kimura, M.; Ishii, H.; Kusaka, M.; Kaizu, K.; Fuji, A. Joining phenomena and fracture load of friction welded joint between pure aluminium and low carbon steel. *Sci. Technol. Weld. Join.* **2009**, *14*, 388–395. [CrossRef]

14. Prater, T. Solid-state joining of metal matrix composites: A survey of challenges and potential solutions. *Mater. Manuf. Proc.* **2011**, *26*, 636–648. [CrossRef]

15. Lean, P.P.; Gil, L.; Ureña, A. Dissimilar welds between unreinforced AA6082 and AA6092/SiC/25p composite by pulsed-MIG arc welding using reinforced filler alloys (Al-5Mg and Al-Si). *J. Mater. Proc. Technol.* **2003**, *143*, 846–850. [CrossRef]

16. Ureña, A.; Escalera, M.D.; Gil, L. Influence of interface reactions on fracture mechanisms in TIG arc-welded aluminium matrix composites. *Compos. Sci. Technol.* **2000**, *60*, 613–622. [CrossRef]

17. Qin, G.; Lei, Z.; Su, Y.; Fu, B.; Meng, X.; Lin, S. Large spot laser assisted GMA brazing-fusion welding of aluminium alloy to galvanized steel. *J. Mater. Proc. Technol.* **2014**, *214*, 2684–2692. [CrossRef]

18. Wielage, B.; Klose, H. Das Aluminiumlöten von Wärmetauschern. *DVS-Rep.* **1995**, *166*, 88–90.

19. Wielage, B.; Martinez, L. Aluminiumlöten bei 550 °C—Eigenschaften von ZnAl-Verbindungen. *DVS-Rep.* **2001**, *212*, 214–217.

20. Wielage, B.; Trommer, F. Löten von Aluminium mit Zinkbasisloten. *Schweiß. Schneid.* **2003**, *5*, 273–275.

21. Song, J.L.; Lin, S.B. Effects of Si additions on intermetallic compound layer of aluminum-steel TIG welding-brazing joint. *J. Alloys Comp.* **2009**, *488*, 217–222. [CrossRef]

22. Elßner, M.; Weis, S.; Grund, T.; Hausner, S.; Wielage, B.; Wagner, G. Lichtbogenlöten von Aluminiummatrix-Verbundwerkstoffen mit AlAgCu-Loten. *Werkstoffe und werkstofftechnische Anwendungen* **2014**, *52*, 181–186.

23. Elßner, M.; Weis, S.; Wagner, G. Joining of Aluminum Matrix Composites and Stainless Steel by Arc Brazing. *Mater. Sci. Forum* **2015**, *825*, 393–400. [CrossRef]

24. Dorn, L. *Hartlöten und Hochtemperaturlöten*; Expert Verlag: Renningen, Germany, 2007.

25. Yuan, Y.; Lee, T.R. *Contact Angle and Wetting Properties*; Springer: Heidelberg/Berlin, Germany, 2013.

26. Zaremba, P. *Hart- und Hochtemperaturlöten*; Verlag fuer Schweisstechnik DVS-Verlag: Duesseldorf, Germany, 1988.

27. N.N. *Brazing Handbook*; American Welding Society: Miami, FL, USA, 1991.

28. Müller, W.; Müller, J.-U. *Löttechnik—Leitfaden für die Praxis*; Verlag fuer Schweisstechnik DVS-Verlag: Duesseldorf, Germany, 1995.

29. Weis, S.; Elßner, M.; Wielage, B.; Wagner, G. Wetting behavior of AlAgCu brazing filler on aluminum matrix composites and stainless steel. *Weld. World* **2017**, *61*, 383–389. [CrossRef]

30. Liu, S.; Zhao, S.; Zhang, Q. Phase Diagram of the Aluminium-Copper-SilverAlloy System. *Acta Metall. Sin.* **1983**, *19*, 70–73.

31. Witusiewicz, V.T.; Hecht, U.; Fries, S.G.; Rex, S. The Ag–Al–Cu system: II. A thermodynamic evaluation of the ternary system. *J. Alloys Comp.* **2005**, *387*, 217–227. [CrossRef]

32. Schmid, G.; Wetter, H. *Stanzniet- und Hybridfügeverfahren im Karosseriebau*; Konferenzband Join-Tec: Halle, Germany, 2005; pp. 42–57.

33. Data Sheet 1.4301. Available online: http://www.ll-adelsdorf.com/pdfdoku/Edelstahl/1.4301.pdf (accessed on 13 June 2014).

34. Springer, H.; Kostka, A.; Dos Santos, J.F.; Raabe, D. Influence of intermetallic phases and Kirkendall-porosity on the mechanical properties of joints between steel and aluminium alloys. *Mater. Sci. Eng. A* **2011**, *528*, 4630–4642. [CrossRef]

35. Elßner, M.; Weis, S.; Wagner, G.; Grund, T. Joining of material compounds of aluminium matrix composites (AMC) by arc brazing using Al-Ag-Cu system filler alloy. *Weld. World* **2017**, *61*, 405–411. [CrossRef]

36. Potesser, M.; Schoeberl, T.; Antrekowitsch, H.; Bruckner, J. The Characterization of the Intermetallic Fe-Al Layer of Steel-Aluminum Weldings. *EPD Congr.* **2006**, *1*, 167.

metals MDPI

Article

The Effect of Interlayer Materials on the Joint Properties of Diffusion-Bonded Aluminium and Magnesium

Stefan Habisch [1],*, Marcus Böhme [2], Siegfried Peter [3], Thomas Grund [2] and Peter Mayr [1]

[1]   Institute of Joining and Assembly, Chemnitz University of Technology, D-09107 Chemnitz, Germany; peter.mayr@mb.tu-chemnitz.de

[2]   Institute of Materials Science and Engineering, Chemnitz University of Technology, D-09107 Chemnitz, Germany; marcus.boehme@mb.tu-chemnitz.de (M.B.); thomas.grund@mb.tu-chemnitz.de (T.G.)

[3]   Institute of Physics, Chemnitz University of Technology, D-09107 Chemnitz, Germany; s.peter@physik.tu-chemnitz.de

*   Correspondence: stefan.habisch@mb.tu-chemnitz.de; Tel.: +49-371-53132336

Received: 15 December 2017; Accepted: 12 February 2018; Published: 17 February 2018

Abstract: Diffusion bonding is a well-known technology for a wide range of advanced joining applications, due to the possibility of bonding different materials within a defined temperature-time-contact pressure regime in solid state. For this study, aluminium alloys AA 6060, AA 6082, AA 7020, AA 7075 and magnesium alloy AZ 31 B are used to produce dissimilar metal joints. Titanium and silver were investigated as interlayer materials. SEM and EDXS-analysis, micro-hardness measurements and tensile testing were carried out to examine the influence of the interlayers on the diffusion zone microstructures and to characterize the joint properties. The results showed that the highest joint strength of 48 N/mm^2 was reached using an aluminium alloy of the 6000 series with a titanium interlayer. For both interlayer materials, intermetallic Al-Mg compounds were still formed, but the width and the level of hardness across the diffusion zone was significantly reduced compared to Al-Mg joints without interlayer.

Keywords: diffusion bonding; aluminum; magnesium; interlayer; titanium; silver; PVD-coating; microstructure; tensile testing; fracture surface analysis

---

1. Introduction

Diffusion bonding is a widely used technology for creating similar and dissimilar joints from challenging materials. These materials are usually joined at elevated temperatures, between 0.5 and 0.8 of the absolute melting point, using a defined contact pressure with a joining time ranging from a few minutes to a few hours. However, the joining process and weldability are highly influenced by the tendency to form oxide layers, e.g., for aluminium and magnesium [1,2]. To improve the limited weldability of aluminium and magnesium, surface treatment is necessary to remove the stable and tenacious oxide layer and limit the formation of brittle intermetallic Al-Mg compounds [3,4].

Diffusion bonding of aluminium and magnesium with interlayer materials has been extensively investigated to improve the joint properties and promote atomic diffusion processes across the bond line [1,3–10]. For this purpose, a large group of interlayer metals is available to influence the properties of the diffusion zone.

Nickel with a thickness of 6 µm was used as interlayer for diffusion bonding of pure aluminium and pure magnesium. After mechanical surface treatment by grinding, both materials were bonded at 440 °C, 1 N/mm^2 contact pressure and 60 min bonding time. The results showed that the formation of the intermetallic Al-Mg compound can be avoided, but at the interfaces of both aluminium/nickel and nickel/magnesium, brittle intermetallic compounds were formed. Therefore, a relatively low joint strength of 26 N/mm^2 was achieved [5].

Good results were achieved for a diffusion bond between AA 6061 and AZ 31 B with zinc or a zinc-aluminium interlayer with a thickness of 35 μm. A shear strength of 86 N/mm^2 was reached, due to the eutectic formation of Al-Zn compounds along the bond line. No formation of intermetallic Al-Mg compounds was observed, but intermetallic Mg-Zn compounds still limited the joint strength [6,7].

Silver as interlayer material was examined for diffusion bonds of pure aluminium to pure magnesium. The maximum joint strength of 14.5 N/mm^2 was reached at 390 °C bonding temperature, 30 min holding time and 5 N/mm^2 contact pressure with a silver layer thickness of 25 μm. The silver diffused completely into both base materials, and brittle intermetallic compounds of Al-Mg and Mg-Ag were formed. Furthermore, the hardness measurement across the diffusion zone revealed a hardness peak on the magnesium side of the diffusion zone, which consisted of different Mg-Ag phases and weakened the joint [8,9].

The aim of this investigation was to systematically study the effect of titanium and silver as interlayer materials on microstructure and mechanical properties for different Al-Mg diffusion bonds.

2. Experimental Section

2.1. Materials, Equipment and Process Parameters

For this study, the aluminium alloys AA 6060, AA 6082, AA 7020 and AA 7075 and the magnesium alloy AZ 31 B were investigated. The chemical composition of the materials is summarized in Table 1. All aluminium alloys were used in the T6 heat-treated condition except for AA 6060, which was used in condition T66.

Table 1. Chemical composition of the base materials according to manufacturer's specifications [11,12].

Materials	Chemical Composition in wt %								
	Al	Mg	Zn	Si	Cu	Mn	Fe	Ti	Cr
AA 6060	bal.	0.35–0.6	0.15	0.3–0.6	0.1	0.1	0.1–0.3	0.1	0.05
AA 6082	bal.	0.6–1.2	0.2	0.7–1.3	0.1	0.4–1.0	0.5	0.1	0.25
AA 7020	bal.	1.0–1.4	4.0–5.0	0.35	0.2	0.05–0.5	0.4	-	0.1–0.35
AA 7075	bal.	2.1–2.9	5.1–6.1	0.4	1.2–2.0	0.3	0.5	0.2	0.18–0.28
AZ 31 B	2.8–3.0	bal.	1.0	-	-	-	-	-	-

Cylindrical specimens 20 mm in diameter and 15 mm in length for the aluminium alloys and 10 mm in length for the magnesium alloy were machined. Aluminium samples were produced by turning. The magnesium samples were cut out of sheet metal using a water jet. The joining surfaces were therefore equivalent to the rolled surfaces of the sheets.

All surfaces were characterized by tactile surface measurement (Hommel Etamic T8000, Jenoptic, Jena, Germany). An area of 1.5 mm by 1.5 mm was measured with a tip radius of 5 μm and 90° angle (measuring sensor: Hommel Etamic TKL 100, Jenoptic, Jena, Germany) and a measuring speed of 0.15 mm/s.

Joining was carried out using a PVA TePla COV 323 HP diffusion bonding machine (PVA TePla AG, Wettenberg, Germany) directly coupled to a glovebox. Details of the setup can be seen in Figure 1 and in [13]. Bonding temperatures of max. 1100 °C and loads of max. 100 N/mm^2 for an area of 300 mm by 300 mm. Diffusion bonding experiments were carried out under argon with a pressure of 700 hPa. A bonding temperature of 415 °C, bonding time of 60 min and a contact pressure of 8.5 N/mm^2 was used, based on preliminary experiments on aluminium and magnesium diffusion bonding. All bonding parameters, such as temperature, time and pressure, were automatically controlled throughout the whole bonding process. For each configuration, six specimens were bonded. One specimen was subjected to a detailed metallographic investigation and micro hardness measurements and five specimens were used for tensile testing. The bonded samples were machined to top hat samples (Figure 2a), in accordance with the ram tensile test [14]. The tensile strength was determined using a universal test machine and a special punch-die-equipment (Figure 2b). Fracture surfaces of tensile specimens and the microstructural characteristics of the diffusion zone were examined using a SEM (Zeiss LEO1455VP, Carl Zeiss AG, Oberkochen, Germany) and EDXS analysis (EDAX Genesis, AMETEK, Inc., Berwyn, PA, USA).

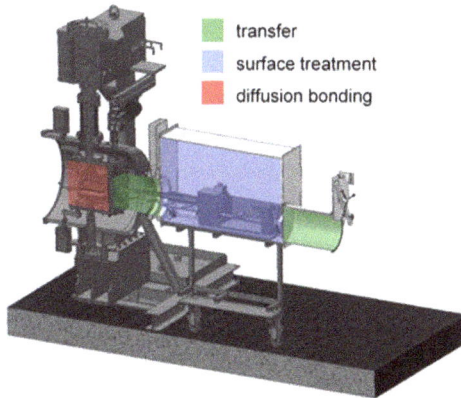

Figure 1. Sectional view of the technology for in-line surface treatment and diffusion bonding.

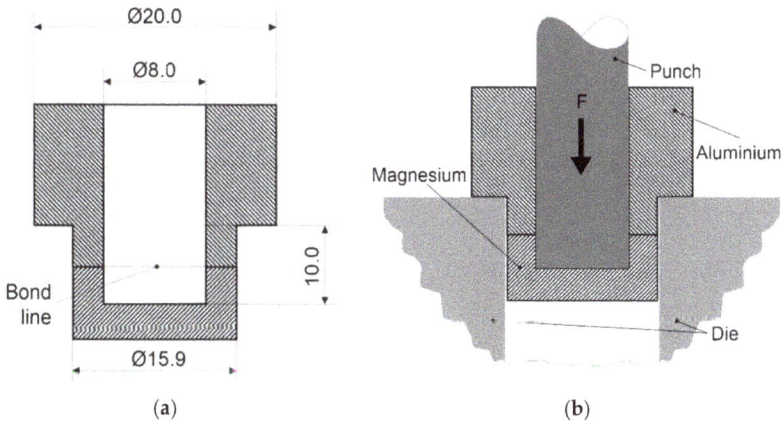

Figure 2. Tensile testing of diffusion bonded joints (**a**) characteristic geometrical dimensions of a top hat sample and (**b**) schematic testing setup.

2.2. Surface Preparation

All aluminium samples were machined to a pre-defined surface roughness. Subsequently, the joining surfaces were cleaned using an electron beam and immediately coated with titanium or silver using physical vapor deposition (PVD). The layer thickness for all layers was 2 μm. Therefore, the new formation of oxide layers on the joining surfaces was inhibited. Titanium was chosen as interlayer because the interlayer will form a diffusion barrier for the intermetallic Al-Mg compounds during diffusion bonding. However, the interdiffusion of Al-Ti and Mg-Ti will form the joint between the aluminum and magnesium materials. In contrast to titanium, silver shows a good diffusivity of Al-Ag and Mg-Ag. The diffusion of silver into the bas materials can influence the interface properties most-likely. Furthermore, the soft silver interlayer will improve the forming process of the joining surfaces to each other.

The magnesium surfaces were chemically treated in the glovebox directly before joining. The oxide layer was removed by immersing the specimens in HNO_3 with a chemical concentration of 10 wt % for 30 s. Both the titanium and silver coatings and the chemical surface treatment were carried out so that the roughness of the joining surfaces was not significantly affected.

2.3. Diffusion Bonding Procedure

After the chemical surface treatment, the aluminium and magnesium samples were assembled on top of each other and transferred from the glovebox into the diffusion bonding chamber. A small contact pressure of ca. 3 N/mm^2 was applied during heating up to 350 °C (heating rate of 3–5 K min^{-1}), to reduce the macroscopic deformation of the magnesium sample. When a temperature of 350 °C was reached, the contact pressure was increased to 8.5 N/mm^2 to reach a near full-faced contact. At 415 °C, the pre-determined bonding time of 1 h was held. At the end of the diffusion bonding procedure, specimens were cooled to room temperature in the chamber at 2–10 K min^{-1}.

3. Results and Discussion

3.1. Joining Surfaces

The surface roughness Sz after turning of the aluminum specimens was $Sz = 5.2 \ \mu m \pm 0.3 \ \mu m$. Coating only slightly changed the joining surfaces to $Sz = 5.1 \ \mu m \pm 0.2 \ \mu m$. The macroscopic characteristic traces of the turning process were still visible (Figure 3a). The chemical surface treatment of AZ 31 B changed the surface roughness from $Sz = 21.6 \ \mu m$ to $Sz = 20.4 \ \mu m$, due to the removal of material (Figure 3b).

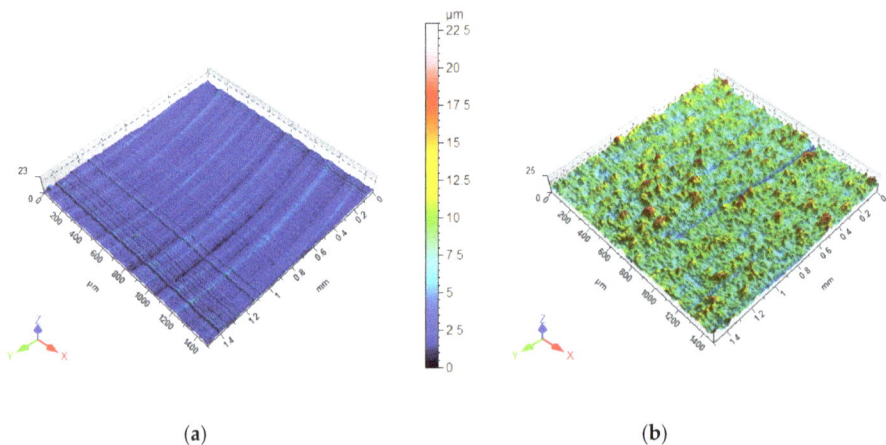

(a) (b)

Figure 3. Tactile measured surface topographies of the joining surfaces after surface treatment with (a) $Sz = 5.2 \ \mu m$ of AA 6082 coated with titanium and (b) $Sz = 20.4 \ \mu m$ of AZ 31 B after chemical treatment.

The rolled magnesium surfaces exhibit a higher surface roughness compared to the aluminium surfaces, but through the enhanced formability of AZ 31 B above 230 °C, a full-faced contact can be assumed. Furthermore, through a higher surface deformation, the amount of lattice defects increases, and thus diffusivity is enhanced.

3.2. Joint Properties with Titanium Interlayer

All aluminium alloys showed a similar morphology in the diffusion zone. Figure 4 depicts the diffusion zone of an AA 7020-AZ 31 B bond. No diffusion of titanium into the aluminium was detected. The interlayer showed similar thickness to the as-coated condition with the exception that a few discontinuities due to fracture of the layer were observed. The EDXS line scan confirmed the interdiffusion of Al and Mg for all joints. Mg, in particular, diffused from the AZ 31 B into the base materials of the 6000 series aluminum alloys, and Al-Mg compounds were formed. The aluminium alloys of the 7000 series showed a higher concentration of Zn in the diffusion zone and intermetallic Al-Mg compounds, respectively (Figure 5). Characterization confirmed that the formation of the intermetallic Al-Mg compound was not avoided by the titanium interlayer. The discontinuous layer structure and the relatively thin layer thickness most likely were not sufficient to suppress interdiffusion. The thickness of ca. 40–50 μm of the intermetallic Al-Mg compound along the interface is similar to the results of joining experiments without interlayer material for the same bonding parameters, e.g., compared to the results of Dietrich et al. [15]. However, no joints with titanium interlayer exhibited cracks or voids along the bond interface, indicating that full metallic continuity had been reached.

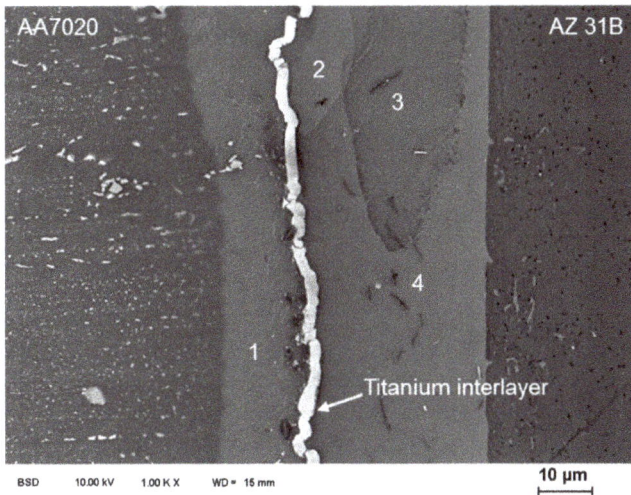

Figure 4. SEM image of the diffusion zone of Al-Mg joint consisting of (**1**) Al_3Mg_2; (**2**) $Al_{12}Mg_{17}$ with ca. 3 at-% Zn; (**3**) $Al_{12}Mg_{17}$ with ca. 2 at-% Zn; and (**4**) $Al_{12}Mg_{17}$.

Figure 6 shows the results of Vickers micro-hardness measurements across the interface of the Al-Mg joints. A significant increase in hardness can be seen for all diffusion zones of the bonds. The magnesium side of the 7000 series aluminium joints show a different behavior compared to the 6000 series. The correlation of Figures 5 and 6 confirms that a higher concentration of zinc on the magnesium side leads to higher micro-hardness.

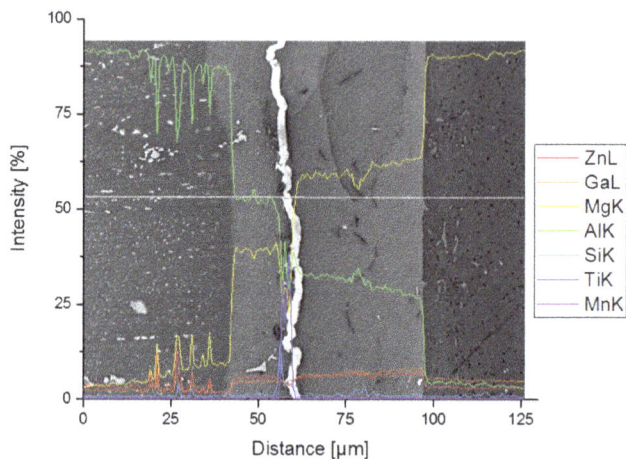

Figure 5. Elemental distribution across the interface of the Al-Mg joint of with titanium interlayer (Figure 4) measured by EDXS line scan.

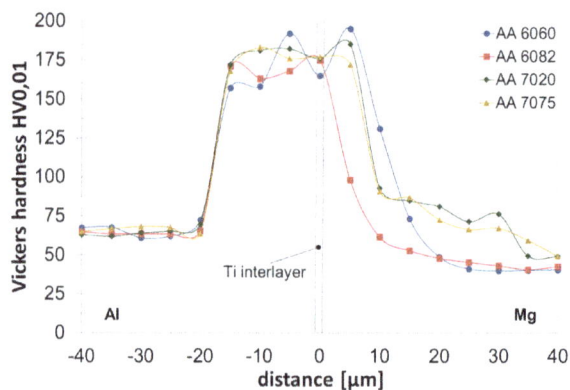

Figure 6. Hardness profiles across the interface of the diffusion-bonded Al-Mg joints with titanium interlayer.

3.3. Joint Properties with Silver Interlayer

In contrast to the results with titanium interlayer, a reaction of the silver coating with the base materials during joining was observed. The silver diffused completely into the base materials, and an intermetallic Al-Mg compound was formed. This zone also contained silver. The white dotted regions in zone 3 of Figure 7 are slightly enriched in silver (Figure 8). Kirkendall voids were formed on the magnesium side, due to the higher coefficient of diffusion of magnesium atoms in aluminium. This is in accordance with other researchers, such as Jafarian et al. [10], who observed similar phenomena for diffusion bonding of aluminium and magnesium with or without interlayer materials. An increase in the contact pressure will avoid the formation of Kirkendall voids; however, macroscopic deformation will increase. The hardness profiles of the Al-Mg joints with silver interlayer exhibit an increase in hardness at the interface of the intermetallic compound $Al_{12}Mg_{17}$ and the magnesium base material. This is pronounced for the aluminum alloys of the 7000 series (Figure 9). The chemical composition of the areas of high hardness again showed an increase of zinc in this area. The enrichment of zinc in the magnesium can form a metallurgical notch in this area away from the bond line. This notch effect is a typical cause of failure for similar and dissimilar joints. For aluminium and magnesium joints without interlayer, the metallurgical notch is located at the interface of the aluminium base material and the Al_3Mg_2 intermetallic compound. These effects were also observed by Liu et al. [7,10].

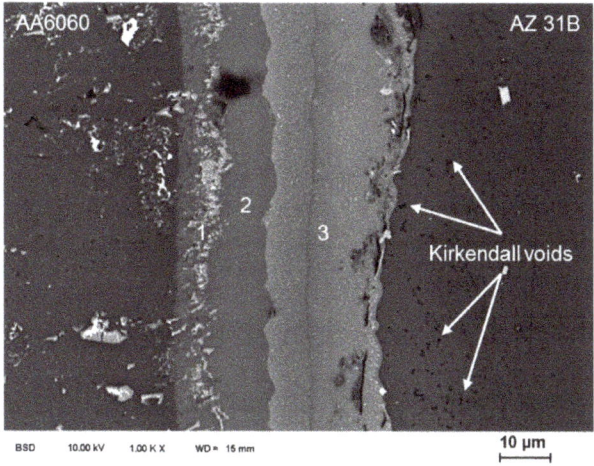

Figure 7. SEM image of the diffusion zone of an Al-Mg joint revealing (**1**) impurities of Al, Mg, Si and O_2, (**2**) the Al-rich intermetallic phase Al_3Mg_2 with ca. 1.5 at-% Ag and (**3**) the Mg-rich intermetallic phase $Al_{12}Mg_{17}$ with ca. 3 at-% Ag.

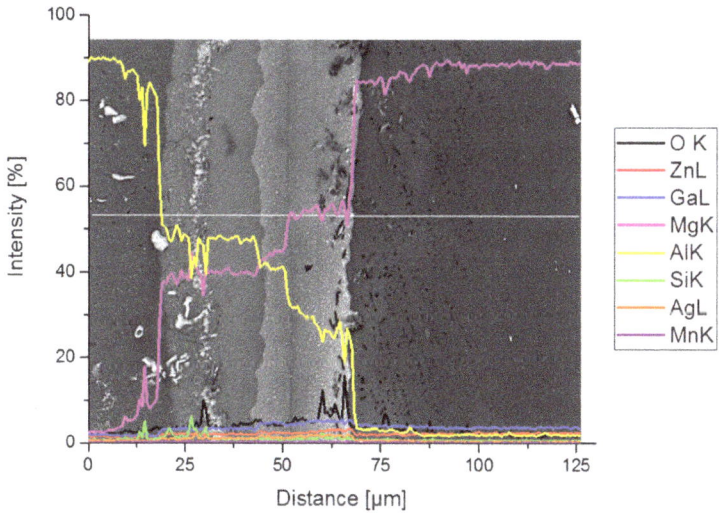

Figure 8. Elemental distribution across the interface of the Al-Mg joint with silver interlayer (Figure 6) measured by EDXS line scan.

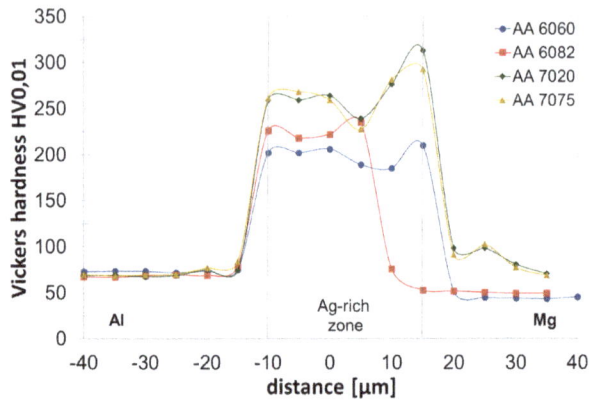

Figure 9. Hardness profiles across the interface of the diffusion-bonded Al-Mg joints with silver interlayer.

3.4. Tensile Testing and Fractography

Figure 10 summarizes the results of the tensile tests with the top hat specimens. Joints with titanium interlayer exhibit the highest strength levels, especially for the aluminium alloys of the 6000 series. The hardness measurements of these show the lowest increase in hardness across the diffusion zone. The hardness profiles of the 7000 series aluminium alloys show an increased hardness by 75 HV 0.01. The fracture surface of an AA 6060 and AZ 31 B joint shows a dense layer of the titanium coating with only few areas of intermetallic Al-Mg compounds (Figure 11). For both the AA 6060 and the AA 6082 aluminium alloys, the fracture surfaces exhibited a dense titanium coating that corresponds to the high rupture strength of 48.3 N/mm^2. However, lower joint strengths were observed for the 7000 series joints with titanium interlayer. The fracture surfaces of the AA 7020 reveals a higher amount of intermetallic Al-Mg compounds (Figure 12). These intermetallic compounds exhibit very brittle behavior, especially when exposed to a tensile load. As a consequence, the base material strength of magnesium of 260 N/mm^2 was not nearly reached. Furthermore, a partial delamination of the titanium coating was observed on the fracture surfaces (Figure 12, No. 3). Significant lower joint strengths were achieved using silver as interlayer material. The hardness profiles (Figure 8) already showed both a higher averaged hardness level in the diffusion zones and a hardness peak on the magnesium side of the 7000 series aluminium alloys. The fracture surface of the aluminium side, shown in Figure 13, exhibits a rough surface, which is characteristic of brittle failures of intermetallic compounds. The EDXS analysis of the fracture surfaces showed that Al$_3$Mg$_2$ and Al$_{12}$Mg$_{17}$ intermetallic phases are present. Despite high hardness on the magnesium side and the confirmed metallurgical notch in this position, the joint finally failed in-between the intermetallic compounds. As a result, the rupture strength for AA 7020 was only 27.7 N/mm^2. In contrast to the 6000 series aluminium alloys (Figure 13), the fracture surface of AA 7020 (Figure 14) appears smoother, but still contains rough areas. The fracture location was probably between the aluminium base material and the intermetallic Al-Mg compound, which contained traces of silver and zinc. Similar fractures as for AA 6060 were observed for AA 6082 and AA 7075.

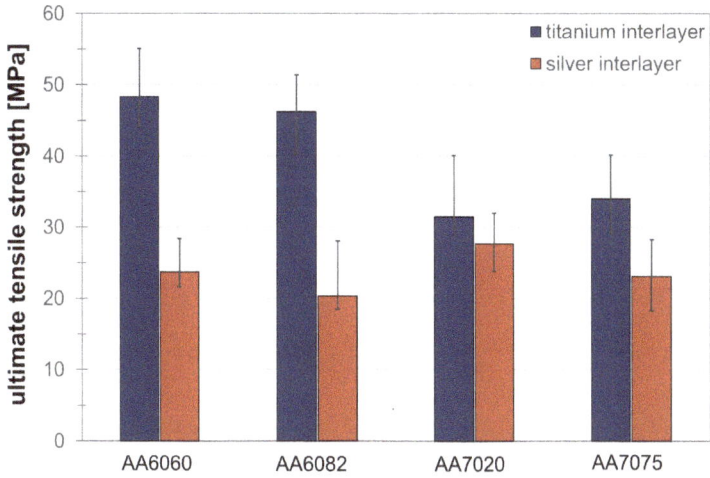

Figure 10. Joint strength of diffusion-bonded aluminum and magnesium with titanium or silver interlayers.

Figure 11. SEM image of a fracture surface of an AA 6060 and AZ 31B joint with titanium interlayer on the aluminium side revealing (**1**) the surface of the titanium coating with only a few areas of (**2**) intermetallic Al-Mg compounds.

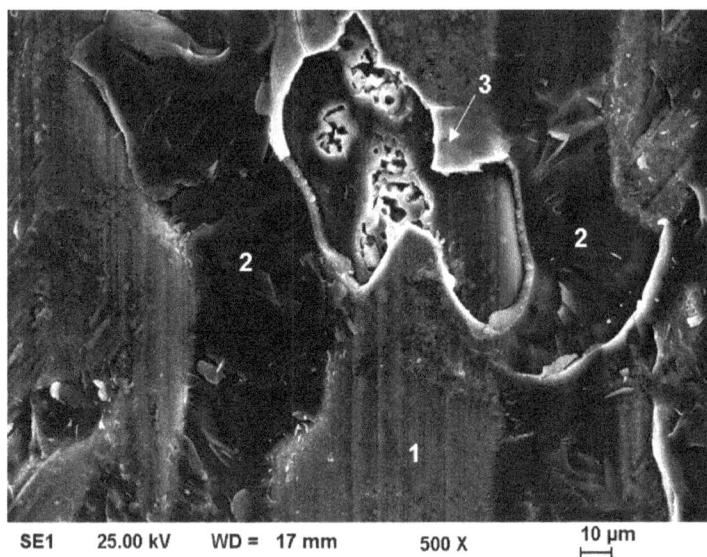

Figure 12. SEM image of a fracture surface of a AA 7020 and AZ 31B joint with titanium interlayer on the aluminium side and (**1**) the surface of the titanium coating, (**2**) intermetallic Al-Mg compounds and (**3**) a delamination of the coating from the aluminium surface.

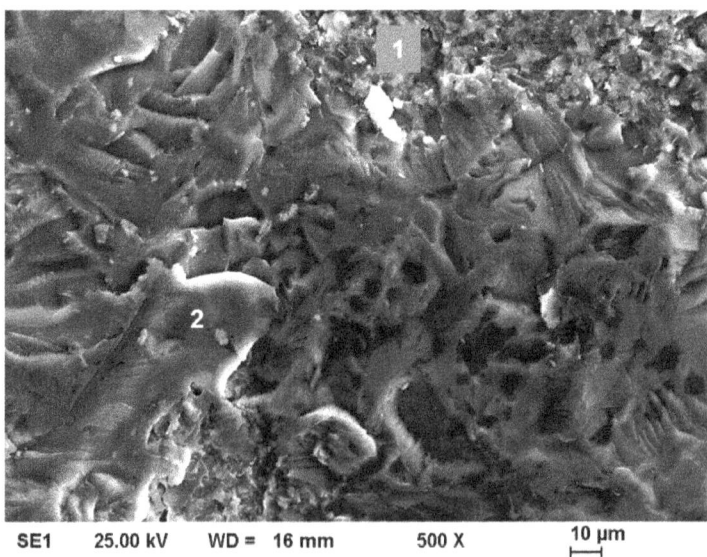

Figure 13. SEM image of a fracture surface of a AA 6060 and AZ 31B joint with silver interlayer on the aluminum side (**1**) presumably Al_3Mg_2 containing traces of silver and (**2**) presumably $Al_{12}Mg_{17}$ containing traces of silver.

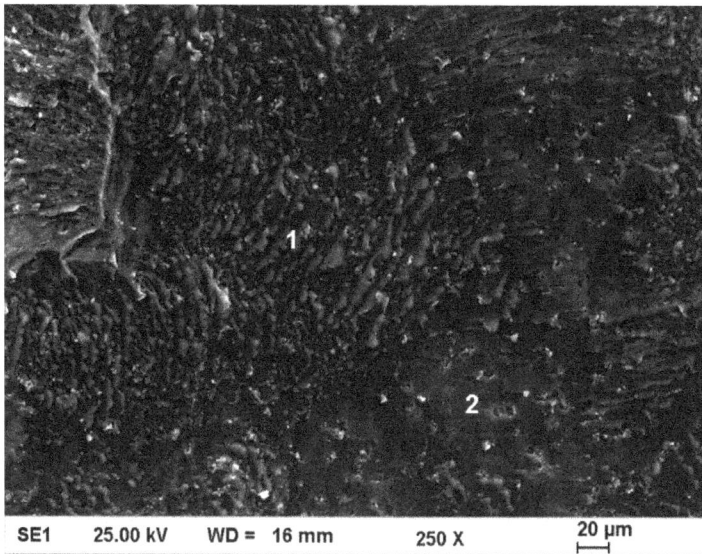

Figure 14. SEM image of a fracture surface of a AA 7020 and AZ 31B joint with silver interlayer on the aluminium side (**1**) presumably Al_3Mg_2 containing traces of silver and zinc and (**2**) aluminium base material with traces of magnesium.

4. Conclusions

In the present study, the effect of titanium and silver as interlayer materials on diffusion-bonded AA 6060, AA 6082, AA 7020, AA 7075 and AZ 31 B dissimilar joints was examined. The following conclusions can be derived:

Titanium and silver as interlayers, deposited by a PVD-process on the oxide-free aluminum surface, made redundant a further surface treatment after exposure of specimens to air.

A solid and almost dense interlayer was formed by titanium on the surface. However, intermetallic Al-Mg compounds were still formed during diffusion bonding. Nevertheless, the hardness measurements across the interface showed a lower hardness level for the diffusion zone compared to Al-Mg joints without interlayers. As a consequence, a joint strength of up to 48 N/mm^2 was reached.

The 2 µm thick silver interlayer completely diffused into both base materials and formed a brittle diffusion zone. Kirkendall voids were observed on the magnesium side and a joint strength of only 28 N/mm^2 was achieved.

Fracture of the joints with titanium interlayer occurred at the interface of the intermetallic Al-Mg compound and the titanium. As the silver diffused into the base materials, the joints failed at the interface of Al_3Mg_2 and $Al_{12}Mg_{17}$.

Acknowledgments: The authors gratefully acknowledge the funding by the German Research Foundation (Deutsche Forschungsgemeinschaft, DFG) within the framework of the Collaborative Research Centre 692 (SFB HALS 692).

Author Contributions: Stefan Habisch carried out all joining experiments and is the main author of this manuscript. Marcus Böhme supported the project with microstructural characterization using SEM and EDS. Siegfried Peter supported the project with the PVD coatings of titanium and silver. Thomas Grund and Peter Mayr are senior scientists, developed and supervised the project "Joining concepts for bulk and sheet metal structures of high-strength light-weight materials" within SFB HALS 692.

Conflicts of Interest: The authors declare no conflict of interest.

References

1. Shirzadi, A.A.; Assadi, H.; Wallach, E.R. Interface evolution and bond strength when diffusion bonding materials with stable oxide films. *Surf. Interface Anal.* **2001**, *31*, 609–618. [CrossRef]
2. Huang, Y.; Humphreys, F.J.; Ridley, N.; Wang, C. Diffusion bonding of hot rolled 7075 aluminium alloy. *Mater. Sci. Technol.* **1998**, *14*, 405–410. [CrossRef]
3. Cherepy, N.J.; Shen, T.H.; Esposito, A.P.; Tillotson, T.M. Characterization of an effective cleaning procedure for aluminum alloys: Surface enhanced Raman spectroscopy and zeta potential analysis. *J. Colloid Interface Sci.* **2005**, *282*, 80–86. [CrossRef] [PubMed]
4. Saleh, H.; Reichelt, S.; Schmidtchen, M.; Schwarz, F.; Kawalla, R.; Krüger, L. Effect of inter-metallic phases on the bonding strength and forming properties of Al/Mg sandwiched composite. *Key Eng. Mater.* **2014**, *622*, 467–475. [CrossRef]
5. Zhang, J.; Luo, G.; Wang, Y.; Shen, Q.; Zhang, L. An investigation on diffusion bonding of aluminium and magnesium using a Ni interlayer. *Mater. Lett.* **2012**, *83*, 189–191. [CrossRef]
6. Liu, L.M.; Zhao, L.M.; Xu, R.Z. Effect of interlayer composition on the microstructure and strength of diffusion bonded Mg/Al joint. *Mater. Des.* **2009**, *30*, 4548–4551. [CrossRef]
7. Liu, L.M.; Zhao, L.M.; Wu, Z.H. Influence of holding time on microstructure and shear strength of Mg–Al alloys joint diffusion bonded with Zn–5Al interlayer. *Mater. Sci. Technol.* **2011**, *27*, 1372–1376. [CrossRef]
8. Wang, Y.; Luo, Q.; Shen, Q.; Wang, C.; Zhang, L. Effect of Holding Time on Microstructure and Mechanical Properties of Diffusion-Bonded Mg1/Pure Ag Foil/1060Al Joints. *Key Eng. Mater.* **2014**, *616*, 280–285. [CrossRef]
9. Wang, Y.; Luo, Q.; Zhang, J.; Shen, Q.; Zhang, L. Microstructure and mechanical properties of diffusion-bonded Mg–Al joints using silver film as interlayer. *Mater. Sci. Eng. A* **2013**, *559*, 868–874. [CrossRef]
10. Jafarian, M.; Khodabandeh, A.; Manafi, S. Evaluation of diffusion welding of 6061 aluminum and AZ31 magnesium alloys without using an interlayer. *Mater. Des.* **2015**, *65*, 160–164. [CrossRef]
11. Magnesium—Otto Fuchs GmbH. Available online: https://www.otto-fuchs.com/en/competence/materials/magnesium.html (accessed on 29 November 2017).
12. Bikar Aluminium GmbH. Available online: http://www.bikar.com/aluminium-round-bars.html (accessed on 19 January 2018).
13. Mayr, P.; Habisch, S.; Haelsig, A.; Georgi, W. Challenges of joining lightweight materials for dissimilar joints. In Proceedings of the 10th International Conference on Trends in Welding Research and 9th International Welding Symposium of Japan Welding Society (9WS), Tokyo, Japan, 11–14 October 2016.
14. Zatorski, Z. Evaluation of steel clad plate weldability using ram tensile test method. *Mater. Trans.* **2007**, *3*, 229–238.
15. Dietrich, D.; Nickel, D.; Krause, M. ; Lampke; T.; Coleman, M.P.; Randle, V. Formation of intermetallic phases in diffusion-welded joints of aluminium and magnesium alloys. *J. Mater. Sci.* **2011**, *46*, 357354. [CrossRef]

Article

Finite Element Simulation of the Presta Joining Process for Assembled Camshafts: Application to Aluminum Shafts

Robert Scherzer [1,*], Sebastian Fritsch [2], Ralf Landgraf [1], Jörn Ihlemann [1] and Martin Franz-Xaver Wagner [2]

1   Chair of Solid Mechanics, Institute of Mechanics and Thermodynamics, Chemnitz University of Technology, Reichenhainer Str. 70, D-09126 Chemnitz, Germany; ralf.landgraf@mb.tu-chemnitz.de (R.L.); joern.ihlemann@mb.tu-chemnitz.de (J.I.)
2   Chair of Materials Science, Institute of Materials Science and Engineering, Chemnitz University of Technology, Erfenschlager Str. 73, D-09125 Chemnitz, Germany; sebastian.fritsch@mb.tu-chemnitz.de (S.F.); martin.wagner@mb.tu-chemnitz.de (M.F.-X.W.)
*   Correspondence: robert.scherzer@mb.tu-chemnitz.de; Tel.: +49-371-531-33848

Received: 22 December 2017; Accepted: 6 February 2018; Published: 11 February 2018

Abstract: This work shows a sequence of numerical models for the simulation of the Presta joining process: a well-established industrial process for manufacturing assembled camshafts. The operation is divided into two sub-steps: the rolling of the shaft to widen the cam seat and the joining of the cam onto the shaft. When manufactured, the connection is tested randomly by loading it with a static torque. Subsequently, there are three numerical models using the finite element method. Additionally, a material model of finite strain viscoplasticity with nonlinear kinematic hardening is used throughout the whole simulation process, which allows a realistic representation of the material behavior even for large deformations. In addition, it enables a transfer of the deformation history and of the internal stresses between different submodels. This work also shows the required parameter identification and the associated material tests. After comparing the numerical results with experimental studies of the manufacturing process for steel-steel connections, the models are used to extend the joining process to the utilization of aluminum shafts.

Keywords: numerical modeling; finite element method; large deformations; plasticity; aluminum; compression tests; torsion tests; shaft-hub connection; camshaft

---

1. Introduction

In materials science and continuum mechanics, research is done on the development of complex constitutive models that are able to reproduce the elastic-viscoplastic material behavior of metals at large deformations. Nonetheless, there still is a lack of usage of those models in industrial applications, where detailed and complex simulation models are necessary, for example to analyze production processes and the resulting behavior of the produced structures. Moreover, increasing requirements of being lightweight and having durability have to be taken into account. Thus, numerically-efficient implementations of material models with high prediction accuracy and applicability in multi-stage forming processes are required. In this contribution, a phenomenological material model of finite strain viscoplasticity with nonlinear kinematic hardening introduced by Shutov and Kreißig [1] is applied to an industrial multi-stage manufacturing process: the Presta joining process (PJP).

PJP is the world's leading process in the series production of assembled camshafts [2]. Compared to conventional cast or forged camshafts, the process allows up to 30% reduced weight. This causes a significant inertia reduction of the rotating camshafts [2,3]. During the multi-step manufacturing process, a shaft-hub connection between a shaft and cams is created. In the first step, a local widening of the shaft at the latter cam seat position is created by forming it with rollers with an approximately sinusoidal cross-sectional profile. Due to the rotation of the shaft between three rollers, a circumferentially-oriented profile is formed (Figure 1a). On the shaft, the peaks of the resulting profile have an increasing diameter with respect to the original diameter. In the second sub-step, a cam with an inner profile oriented parallel to the shaft axis is forced onto the widened cam seat (Figure 1b). Thus, the inner profile of the cam and the circumferentially-oriented profile of the shaft are oriented orthogonally to each other, which causes large local plastic deformations during the joining process and creates a tight press and form fit. In a subsequent quality control step, randomly-selected camshafts are tested within a static torsional test (Figure 1c). Thereby, the maximum transmittable torque of the connection is evaluated.

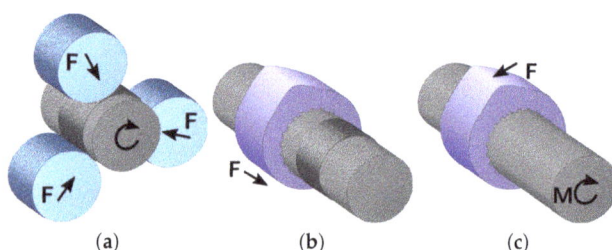

Figure 1. Scheme of the PJP: (**a**) rolling of the shaft to create a local widening; (**b**) the joining process that forces the cam onto the widened cam seat; (**c**) loading of the connection to evaluate the maximum transmittable torque.

The PJP process is well known experimentally, but still has a large development potential. So far, it has only been applied to camshafts consisting of steel-steel connections. Corresponding analyses by the help of the finite element method (FEM) have already been conducted and showed good agreement with the experimentally-obtained results of the different manufacturing steps (see [4–6]). Therein, the application of the phenomenological material model of finite strain viscoplasticity enabled the transfer of the deformation history at the change of simulation models. In this paper, the extension of PJP to the application aluminum shafts is regarded, which provides further weight reduction and in some cases an improved bearing of the camshaft. Thereby, the phenomenological material model of finite strain viscoplasticity is applied to both the steel cam and the aluminum shaft. Experimental tests and the parameter identification procedure to obtain material parameters of the aluminum alloy are presented in this paper. Moreover, the FEM models of the three sub-steps, i.e., rolling, joining and testing by a static torque, are described and applied to the aluminum-shaft/steel-cam connection. Thereby, the producibility of this connection and the final resistance to a static torque loading are analyzed.

2. Materials and Methods

2.1. Material Modeling

The phenomenological material model of finite strain viscoplasticity with nonlinear kinematic hardening introduced by Shutov and Kreißig [1] has been extended to a formulation providing multiple backstresses in Shutov et al. [7] (Figure 2). This material model is able to properly reproduce the material behavior at large deformations [4,7,8].

The formulation used in the numerical models of the joining process contains an elastic region, a von Mises yield criterion, rate-dependent yielding of the Perzyna type, two isotropic hardening mechanisms of the Voce type and two backstresses of the Armstrong–Frederick type. Altogether, the model includes 13 material parameters. This work uses the strategy for parameter identification of Shutov et al. [7]. The shear modulus G, the bulk modulus K and the yield stress σ_F are determined by analyzing the elastic region of the experimentally-determined stress-strain curves or by using standardized table values. Next, the parameters representing the strain rate dependency—the viscosity η and the parameter m of the Perzyna rule—are identified by analyzing the different overstresses that occur at different strain rates (cf. Figure 3b). The remaining eight parameters representing the kinematic ($c_{kin1}, c_{kin2}, \kappa_1, \kappa_2$) and isotropic hardening ($\beta_1, \beta_2, \gamma_1, \gamma_2$) are identified by adapting the material model to experimentally-obtained stress strain curves (cyclic torsion tests on thin hollow tubes; see Section 2.2). The optimization procedure was realized by the help of the nonlinear least-squares solver lsqnonlin provided by MATLAB.

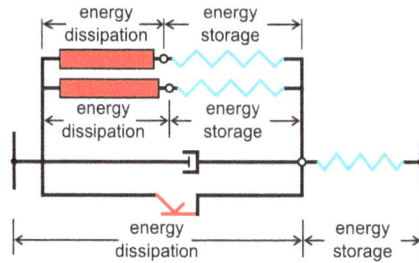

Figure 2. Rheological model of the material model with two backstresses (endochronic Maxwell elements), cf. [7,9].

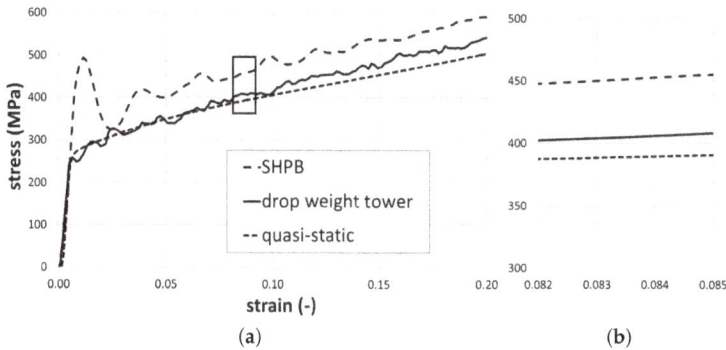

Figure 3. (a) Compression tests of aluminum AA6082 at three different strain rates (quasi-static 10^{-3} s^{-1}, drop weight tower 3×10^2 s^{-1} and SHPB 1.6×10^3 s^{-1}) for the identification of the strain rate dependence; (b) detailed view of the three different stress levels used to calculate the related viscous overstresses.

2.2. Experimental Methods

To experimentally study the mechanical behavior of the aluminum alloy AA6082 (obtained from Bikar-Aluminium GmbH, Korbußen, Germany), cylindrical specimens with a height of 4 mm and a diameter of 4 mm were used in the conventional peak aged condition. The material samples were uniaxially deformed under quasi-static compressive loading conditions at room temperature (RT) with an initial strain rate of 10^{-3} s^{-1} (Figure 3). Additionally, the mechanical behavior under dynamic compression load was characterized. Here, a drop weight tower testing machine was used with a mass of 600 kg. The impact velocity was 1.4 $\frac{m}{s}$, which results in an effective initial strain rate of 3×10^2 s^{-1} (Figure 3). For testing at even higher strain rates, a split-Hopkinson pressure bar (SHPB) setup was used. The SHPB system contains a gas launcher, the striker, incident and transmitter bars and a shock absorber, which are mounted on a flat base. The samples are positioned between the incident and transmitter bars. The gas launcher is charged with a pressurized gas to accelerate the striker bar with an initial impact velocity of $7\,\frac{m}{s}$. This results in an initial strain rate of 1.6×10^3 s^{-1}, which is the same order of magnitude of strain rates like the rolling process (Figure 3). More detailed information on the experimental setups can be taken from previous studies [10,11].

For both drop weight tower and SHPB tests, force data were determined from the strain gauge data recorded on the elastic deforming on the plunger, as well as on the incident and transmitter bars. The deformation of all compression tests was documented with a high-speed camera. To determine the surface strain fields, an optical digital image correlation (DIC) system (Aramis by GOM) was used. The measured displacements and loads were related to initial sample height and cross-section in order to determine engineering strains and stresses, respectively. Then, true (logarithmic) strains and stresses were calculated from the engineering data. For the statistics, three samples were tested for each strain rate.

Furthermore, torsion tests on thin-walled hollow tubes were conducted in a Schenk PTT250 K1 machine. The deformed part of the samples had an outer diameter of 19 mm and an inner diameter of 16 mm. The gauge length was 4 mm. The flow curves were characterized at three different angular velocities: $0.235°$ s^{-1}, $10°$ s^{-1} and $50°$ s^{-1}. In addition, the strain hardening behavior of the materials was analyzed by varying the deformation paths at an angular velocity of $0.235°$ s^{-1}: (i) $+10°/-10°$; (ii) $+10°/-8°$; (iii) $+10°/-5°$.

2.3. Rolling of the Shaft

The rolling of the shaft is performed by three rollers arranged in a rolling tool by an angle of $120°$ (Figure 4). This angle varies slightly over the rolling process by $\pm1°$.

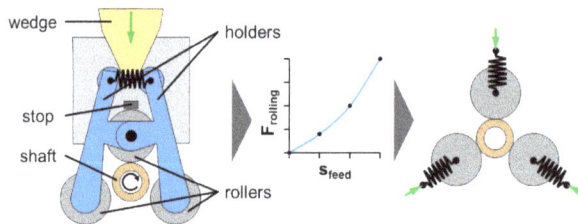

Figure 4. The rolling tool with the two holders, the three rollers and the shaft in the middle; cf. [4].

This arrangement is used in the geometrical model of the rolling process by reducing the volume to a section of 120 degrees and applying cyclically-symmetric boundary conditions. The variation of the rollers' arrangement has not been considered in the model to maintain cyclic symmetry. Additionally, the cam seat is cut into halves in the axial direction. Hence, the geometry of the numerical model includes just a sixth of the shaft geometry (Figure 5). The stiffness of the holders of the rolling tool is represented in the numerical model by multi-linear springs; see Scherzer et al. [4]. Thus, the radial displacement of the rollers is controlled by the stiffnesses of the holders and of the shaft.

The numerical model is calculated using the commercial FEM software ABAQUS 6.14 standard (implicit). The phenomenological material model of finite strain viscoplasticity is implemented by using ABAQUS' user interface for the implementation of material subroutines (UMAT) and employing an efficient implicit integration scheme introduced by Shutov [12]. For the calculation of ABAQUS specific values, the so-called wrapper technology described in Besdo et al. [13] is applied. The model uses linear eight-node brick elements C3D8 with full integration.

The model of the rolling process has been evaluated in Scherzer et al. [4] by comparing experimental and calculated data of the reaction forces of the holders over the whole process with the traditional usage of steel shafts and steel cams. Additionally, the rolled profile has been measured and compared to the computed shaft profile.

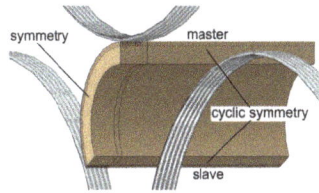

Figure 5. Geometry of the numerical model of the rolling process with axially- and cyclically-symmetric boundary conditions; cf. [4].

2.4. The Joining Process

The inner profile oriented parallel to the shaft axis of the cam is forced onto the widened cam seat in the joining process. This inner profile of the cam requires a fine tangential mesh in the regions of shifting inner diameters. In contrast, the mesh of the shaft in the first simulation step is very coarse compared to its cross-sectional mesh because the resulting shaft profile after the rolling process is almost axisymmetric. Hence, for the second simulation step, a new mesh is required. Additionally, if the outer profile is simplified to a circular contour, it can be reduced to a model including only one tooth of the inner profile (Figure 6a).

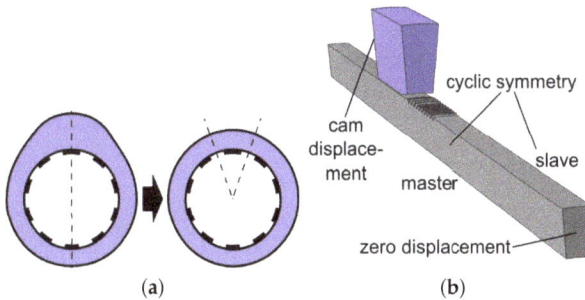

Figure 6. (**a**) Simplification of the cam's outer profile to a circular contour; (**b**) geometry of the numerical model of the joining process reduced to one tooth of the inner profile; cf. [5].

Normally, the change of numerical models together with the change of geometry would cause a loss of information on the internal stresses and the deformation history of the rolling process. Thus, at the start of the joining step, the shaft would be free of stresses and the state variables of the material model would be at the initial state of undeformed material. The variables store the deformation history of the process and include the right Cauchy–Green tensors $\underline{\underline{C}}_i$ and $\underline{\underline{C}}_{ii}$ representing intermediate configurations (see Figure 2), the inelastic arc length s and its dissipative part s_d. However, in Shutov et al. [14], the invariance of the phenomenological material model of finite strain viscoplasticity under an isochoric change of the reference configuration has been introduced. This invariance can be shown by transforming the internal state variables of the material model by $\underline{F}_0$, which is the deformation gradient from the old to the new reference configuration.

Further, this concept has been extended to non-isochoric changes of the reference configuration and applied to the PJP in Scherzer et al. [6]. Hence, the internal state variables at the end of the rolling simulation (i.e., with a certain loading state) are transformed to the undeformed reference configuration at the start of the joining simulation (described by a deformation gradient equal to the unit tensor). As Equation (1) shows, the relative deformation gradient $\underline{F}_0$ required for the transformation equals the deformation gradient of the rolling step $\underline{F}_1$.

$$
\begin{aligned}
\overset{\text{new}}{\underline{C}}_i &= \overline{\left(\underline{F}_1^{-T} \cdot \underline{C}_i \cdot \underline{F}_1^{-1} \right)} & \overset{\text{new}}{S} &= s \\
\overset{\text{new}}{\underline{C}}_{ii} &= \overline{\left(\underline{F}_1^{-T} \cdot \underline{C}_{ii} \cdot \underline{F}_1^{-1} \right)} & \overset{\text{new}}{S}_d &= s_d
\end{aligned}
\qquad \text{with:} \qquad \overline{\underline{X}} = I_3(\underline{X})^{-\frac{1}{3}}\underline{X} \qquad (1)
$$

At the start of every ABAQUS simulation, the deformation gradient also is equal to the unit tensor. Thus, in the first increment of the calculation, the model of the joining step with the new geometry, mesh and a transformed set of internal variables recovers the internal stresses of the rolling process.

Just like the rolling model, the numerical model of the joining step has been evaluated and tested by comparing its results to experimental data of the traditional Presta connection using steel shafts and steel cams. The computed joining force curve is especially in a good agreement with the curves measured during actual series production [5].

2.5. Loading of the Presta Connection

The last step of the simulation package of the PJP is a numerical model for applying a static load in the model to compute the maximum transmittable torque of the connection. In contrast to the model change between the rolling and the joining, the load step uses the same geometrical model and mesh as the previous model. Hence, the numerical model uses the restart option of ABAQUS to continue the calculation after the joining step. This approach automatically maintains the deformation history and internal stresses of the previous steps. Thus, the geometrical entities remain the same as in the joining model (Figure 7).

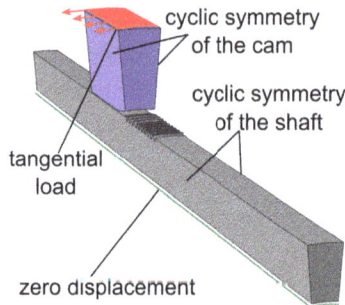

Figure 7. Geometry of the numerical model of the load step with cyclically-symmetric boundary conditions for each of the single parts to enable relative displacement between them.

However, the boundary conditions have to be adapted. The axial position of the cam relative to the shaft is fixed by the friction between the single parts. The tangential twist of the shaft is prevented by a zero displacement boundary condition at its inner surface. The tangential load is applied to the top of the cam and increases linearly until the simulation is aborted because of the cam's displacement (Figure 7).

The cyclic symmetry boundary condition of ABAQUS cannot be applied in this simulation model because of the relative tangential movement between the two parts. ABAQUS allows just one cyclic symmetry boundary condition per model, and this is only applicable to a single master surface (set) and one related slave surface (set). This causes problems when tangential displacements occur. Therefore, the displacements (described in cylindrical coordinates) of every single node at the slave surface become coupled with the corresponding displacement of the master surface. This method has been implemented in the Python script controlling the model.

3. Results

3.1. Material Tests and Parameter Identification for the Aluminum Alloy AA6082

The single numerical models have been developed and verified with the traditional Presta connection of steel cams and steel shafts. Next, the simulation sequence is utilized to achieve the extension of the PJP to the usage of aluminum shafts. To describe the material behavior of AA6082, both literature data and the data given in Section 2.2 are employed. In Section 2.1 it is shown how the material parameters have to be determined. Shutov et al. [7] determined the following numerical parameters: the shear modulus G, the bulk modulus K and the yield stress σ_F (Table 1). Moreover, the viscosity η and the parameter m of the Perzyna rule were identified using the different viscous overstresses obtained by the compression tests (Figure 3 and Table 1).

Table 1. Material parameters G, K and σ_F describing the elastic material behavior and the parameters η and m describing the strain rate dependency of the aluminum alloy AA6082.

G (MPa)	K (MPa)	σ_F (MPa)	η (s)	m (−)
27,300	76,212.5	250	755.48	2.19

Finally, the remaining eight hardening parameters $(c_{kin1}, c_{kin2}, \kappa_1, \kappa_2, \beta_1, \beta_2, \gamma_1, \gamma_2)$ were identified. This was performed in an optimization using MATLAB with the experimental data of the torsion tests (Table 2 and Figures 8 and 9).

Table 2. Material parameters c_{kin1}, c_{kin2}, κ_1, κ_2, β_1, β_2, γ_1 and γ_2 describing the plastic hardening of the aluminum alloy AA6082.

c_{kin1} (MPa)	c_{kin2} (MPa)	κ_1 (−)	κ_2 (−)	β_1 (−)	β_2 (−)	γ_1 (MPa)	γ_2 (MPa)
136.109	51,439.136	0.0717	0.0521	−72.904	31.165	0.000304	2999.53

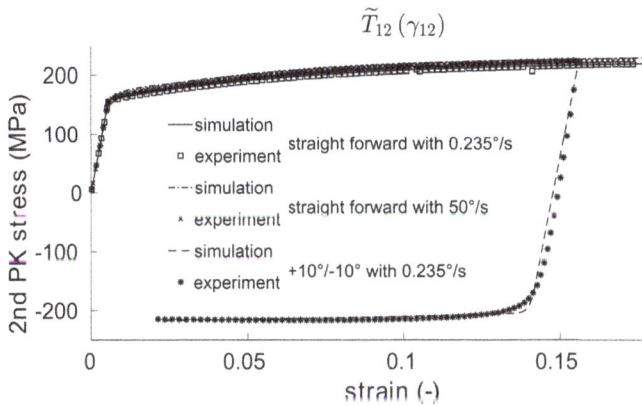

Figure 8. Results of the parameter identification of the hardening effects comparing the calculated stresses (lines) with the stresses from experimental data (dots) of the torsion tests.

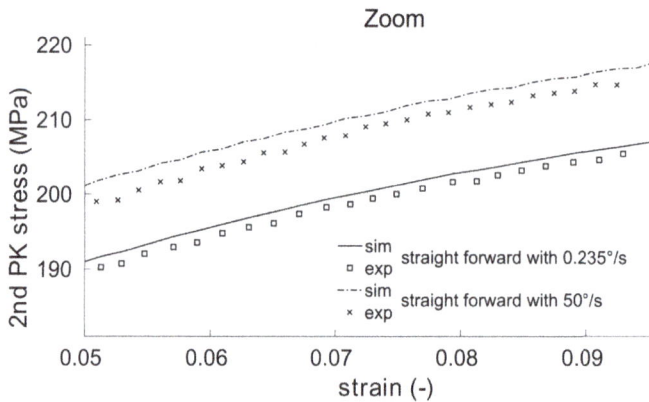

Figure 9. The magnified view of Figure 8 shows the different overstresses at different strain rates.

3.2. Application of the Simulation Sequence on AA6082

Next, the simulation of the rolling process is executed by varying the rollers' cross-sectional profile, the preloading of the holders and the geometric dimensions of the shaft. Here, existing geometrical sets of thyssenkrupp Presta have been used to obtain a fitting arrangement for the AA6082 alloy. For the eventually chosen geometric constraints, the rolling simulation of the AA6082 alloy shows a very good rolling profile. This can be assessed by comparing the resulting shaft profile with the given profile of the roller. Here, the profile is shaped completely without causing plastic deformations at the inner radius of the shaft.

Thus, an adequate local widening has been achieved, which is required for the joining process. As Figure 10 shows, the mesh in the tangential and radial directions is coarse compared to the mesh refinement in the axial direction in order to save computational cost. Fortunately, this partially coarse mesh has a small impact on the convergence of the simulation results since the deformation is approximately axisymmetric and comparably small in the direction of the long edges of the elements.

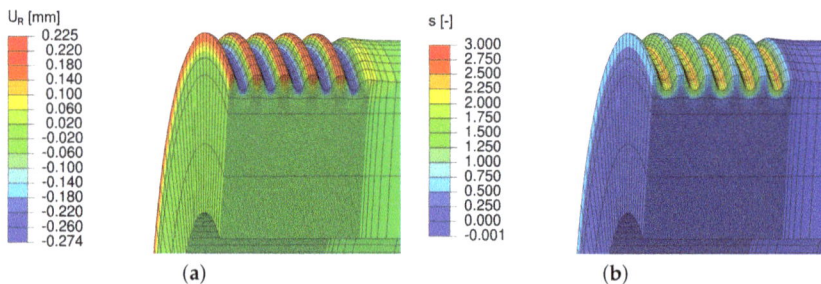

Figure 10. (**a**) Plot of the radial displacement u_r after widening the shaft made from AA6082 through rolling; (**b**) plot of the inelastic arc length s, which can be interpreted as the degree of plastic deformation or as the effective plastic strain.

The phenomenological material model of finite strain viscoplasticity includes the internal state variable s (inelastic arc length) [7], which can be seen as the degree of plastic deformation or as the effective plastic strain. This variable increases monotonically with plastic deformation and independently of the loading path. Thus, it can be used to figure out the zone of plastic deformation (Figure 10b).

Since the internal state variables are transferred at the change of models, the plot of the inelastic arc length s at the start of the joining process in Figure 11a is equivalent to the plot at the end of the rolling step in Figure 10b, although there is a new geometric model and a new mesh. More details on the transformation procedure can be found in Scherzer et al. [6].

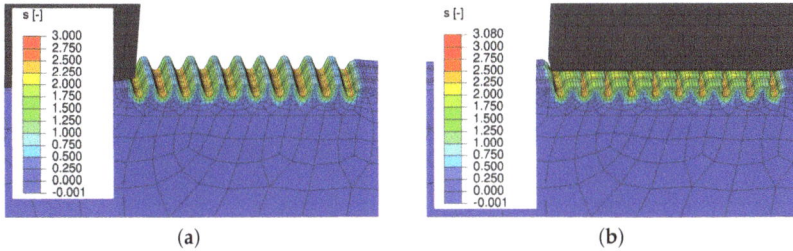

Figure 11. Comparison of the degree of plastic deformation before and after the joining step. (**a**) The transferred inelastic arc length s at the start of the joining step with the new geometry and the new mesh; (**b**) the inelastic arc length s at the end of the joining step with the deformed shaft profile and a cut through the cam. Furthermore, this plot shows the deformed grooves of the cam seat.

After the cam with its inner profile is forced onto the shaft profile, the degree of deformation is increased slightly compared to the initial value of the rolling step (Figure 11b). In the series production, the quality of the connection is measured continuously by monitoring the limits of the joining force (Figure 12a). Besides the form fit due to the plastic deformation in the joining process, there is an additional press fit that results from the elastic compression of the shaft and the elastic decompression of the cam (Figure 12b).

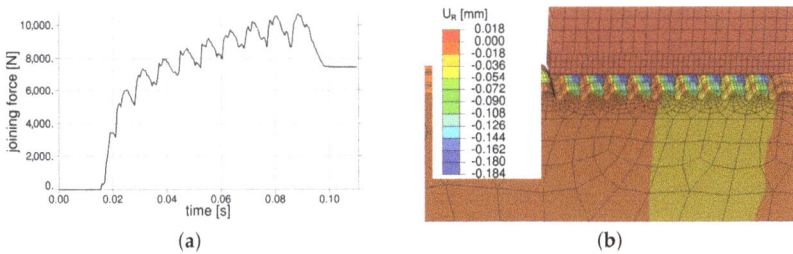

Figure 12. (**a**) Calculated progression of the joining force over time for the joining of the cam onto the aluminum shaft. The single grooves of the rolled profile cause small peaks in the force path. The force grows over the joining process due to the increasing friction surface. (**b**) Radial displacement of the joining step showing the compression of the shaft and the decompression of the cam.

Next, the connection is tested numerically by applying a linear increasing static torque in the loading simulation (cf. Section 2.5). Thereby, the maximum transmittable torque of the connection is determined by evaluating the progression of the cam's tangential displacement over loading time (Figure 13a).

The connection remains intact as long as the displacement increases linearly. In this state, the contact zone between the cam and the shaft is almost zero (Figure 13b (top)). The connection fails as soon as as a critical torque is achieved. There are a further increase of the torque causes plastic deformations in the contact zone, an increasing relative displacement between the cam and the shaft (Figure 13b (bottom)) and a nonlinear progression of the corresponding tangential displacement curve.

The transition point between linear and nonlinear curve shape is employed to identify the critical torque (Figure 13a). Hence, the maximum transmittable torque of the steel-aluminum connection is approximately 115 Nm (Figure 13a).

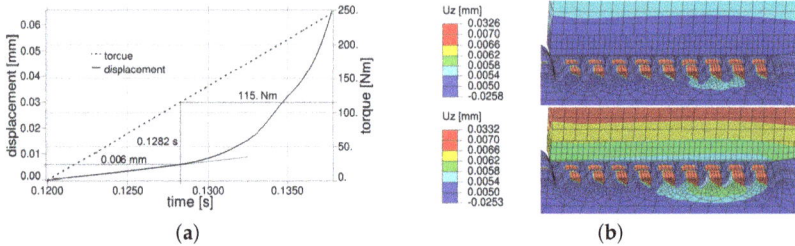

(a)   (b)

Figure 13. (**a**) Linearly-increasing torque with the related tangential displacement of the cam and the determination of the maximum transmittable torque of the connection; (**b**) the tangential displacement of the loading model at $t_0 = 0.128$ s (top) and $t_1 = 0.129$ s (bottom).

4. Discussion

This approach connects three numerical models to a sequence to enable simulations of the PJP. Because of the large deformations that occur during the process, it is necessary to use an appropriate material model. Here, a phenomenological material model of finite strain viscoplasticity has been used. To enable a proper reproduction of the material behavior, it is necessary to perform material tests and an identification of the material parameters. The quality of the resulting set of parameters depends on the complexity of the material tests. For technical reasons, the chosen set of material tests is a combination of torsion and compression tests. Hence, adding more experiments would improve the quality of material parameters.

The rolling of the shaft produces an approximately axisymmetric profile. This means that there is a small gradient of deformation in the tangential direction, and thus, a relatively coarse mesh has been used tangentially. This coarse mesh allows the calculation of the model in an acceptable time span.

The joining process simulation is based on a simplification of the outer contour of the cam. Further investigations with the simulation of the full geometry of the cam could show the influence of this simplification on the joining force. Additionally, localization effects in the deformation of the cam and the rolled profile of the shaft could be investigated.

The maximum transmittable torque obtained in the loading simulation results from a static test of the connection. For the usage in combustion engines there are high demands on the fatigue strength of the connection. The connection shown in this approach most likely does not meet these requirements because it is based on an existing geometrical configuration used for steel-steel connections. An adjustment of the geometrical design, with the aim of improving fatigue behavior, is a key topic of future investigations.

5. Summary

This work combines the simulation models of the rolling of the shaft and of the joining step with a simulation sequence for the joining process and adds a third simulation step for testing the connection by loading it with a static torque. The included usage of the phenomenological material model of finite strain viscoplasticity makes it necessary to perform precise material tests. The underlying experimental methods and the identification of the material parameters have been described.

Finally, the simulation package has been used to test numerically if the usage of aluminum shafts is theoretically possible. Indeed, the numerical results demonstrate the technical feasibility of such an extension of the PJP under the aspect of static loads.

However, considering the application of aluminum shafts in combustion engines, the shafts also have to meet the requirements of fatigue strength. Therefore, several changes in the dimensioning of the shaft, the cam and their attachments have to be made. Moreover, the shaft design shown in this work will be manufactured and tested in the near future. If the experimental tests of this connection verify the computed results, the required design changes can be investigated and implemented.

6. Conclusions

Industrial research often uses built-in material models of commercial FEM software. Possible reasons for this are the required implementations of the models and the expensive material tests to obtain the material parameters. This work proves the concept of the use of a phenomenological material model of finite strain viscoplasticity in an industrial manufacturing process. Additionally, the transfer of the deformation history from the rolling simulation to the joining step has proven the concept of the change of the reference configuration to perform in multi-step analyses.

Altogether, this work has shown the feasibility of the application of aluminum in the PJP by numerical simulations. The next steps of investigation include the validation of numerical results by manufacturing sample camshafts with aluminum shafts, as well as the change of the geometrical design of the shaft and of the cam to meet the requirements of fatigue strength. The simulation sequence shown in this approach may well be an essential help in the process of this development.

Acknowledgments: The research shown in this work originates from the publicly-supported collaborative research center SFB 692 "High-strength aluminum based lightweight materials for safety components", which is supported by the German Research Foundation (DFG). Furthermore, this research is carried out in close cooperation with thyssenkrupp Presta GmbH in Chemnitz. The underlying project is a transfer project with the goal to show possible industrial applications of previous research results of the collaborative research center.

Author Contributions: Sebastian Fritsch designed and performed the experiments and wrote the relating section "Experimental Methods". Robert Scherzer analyzed the data, developed the numerical models, performed the numerical investigations and wrote the remaining parts of the paper. All authors discussed the results and helped with writing the manuscript.

Conflicts of Interest: The authors declare no conflict of interest.

Abbreviations

The following abbreviations are used in this manuscript:

DIC	Digital image correlation
FEM	Finite element method
PJP	Presta joining process
SFB	Sonderforschungsbereich (engl. collaborative research center)
SHPB	Split-Hopkinson pressure bar
UMAT	User subroutine to define a material's mechanical behavior in ABAQUS

References

1. Shutov, A.V.; Kreißig, R. Finite strain viscoplasticity with nonlinear kinematic hardening: Phenomenological modeling and time integration. *Comput. Methods Appl. Mech. Eng.* **2008**, *197*, 2015–2029.
2. Meusburger, P. Lightweight design in engine construction by use of assembled camshafts. *MTZ Worldw.* **2006**, *67*, 10–12.
3. Lengwiler, A. *Fehlerfortpflanzung, Simulation und Optimierung von Prozessketten Anhand der Gebauten Nockenwelle*; Eidgenössische Technische Hochschule ETH Zürich: Zürich, Switzerland, 2011.
4. Scherzer, R.; Silbermann, C.B.; Ihlemann, J. FE-simulation of the Presta joining process for assembled camshafts—Local widening of shafts through rolling. *IOP Conf. Ser. Mater. Sci. Eng.* **2016**, *118*, 12039, doi:10.1088/1757-899X/118/1/012039.
5. Scherzer, R.; Silbermann, C.B.; Landgraf, R.; Ihlemann, J. FE-simulation of the Presta joining process for assembled camshafts—Modelling of the joining process. *IOP Conf. Ser. Mater. Sci. Eng.* **2017**, *181*, 12030, doi:10.1088/1757-899X/181/1/012030.

6. Scherzer, R.; Ihlemann, J. Simulation of the Presta process—Transfer of deformation history. *PAMM* (accepted 2017).

7. Shutov, A.V.; Kuprin, C.; Ihlemann, J.; Wagner, M.X.; Silbermann, C. Experimentelle Untersuchung und numerische Simulation des inkrementellen Umformverhaltens von Stahl 42CrMo4. *Materialwiss. Werkst.* **2010**, *41*, 765–775.

8. Shutov, A.V.; Kreißig, R. Regularized strategies for material parameter identification in the context of finite strain plasticity. *Tech. Mech.* **2010**, *30*, 280–295.

9. Kießling, R.; Landgraf, R.; Scherzer, R.; Ihlemann, J. Introducing the concept of directly connected rheological elements by reviewing rheological models at large strains. *Int. J. Solids Struct.* **2016**, *97*, 650–667.

10. Winter, S.; Schmitz, F.; Clausmeyer, T.; Tekkaya, A.E.; Wagner, M.F.-X. High temperature and dynamic testing of AHSS for an analytical description of the adiabatic cutting process. *IOP Conf. Ser. Mater. Sci. Eng.* **2017**, *181*, 12026, doi:10.1088/1757-899X/181/1/012026.

11. Pouya, M.; Winter, S.; Fritsch, S.; Wagner, M.F.-X. A numerical and experimental study of temperature effects on deformation behavior of carbon steels at high strain rates. *IOP Conf. Ser. Mater. Sci. Eng.* **2017**, *181*, 12022, doi:10.1088/1757-899X/181/1/012022.

12. Shutov, A.V. Efficient implicit integration for finite-strain viscoplasticity with a nested multiplicative split. *Comput. Methods Appl. Mech. Eng.* **2016**, *306*, 151–174.

13. Besdo, D.; Hohl, C.; Ihlemann, J. ABAQUS implementation and simulation results of the MORPH constitutive model. *Const. Models Rubber IV* **2005**, *4*, 223–228.

14. Shutov, A.V.; Pfeiffer, S.; Ihlemann, J. On the simulation of multi-stage forming processes: Invariance under change of the reference configuration. *Materialwiss. Werkst.* **2012**, *43*, 617–625.

Article

Temperature and Particle Size Influence on the High Cycle Fatigue Behavior of the SiC Reinforced 2124 Aluminum Alloy

Lisa Winter *, Kristin Hockauf and Thomas Lampke

Institute of Materials Science and Engineering, Technische Universität Chemnitz, Erfenschlager Str. 73, 09125 Chemnitz, Germany; kristin.hockauf@mb.tu-chemnitz.de (K.H.); thomas.lampke@mb.tu-chemnitz.de (T.L.)
* Correspondence: lisa.winter@mb.tu-chemnitz.de; Tel.: +49-371-531-32632

Received: 11 December 2017; Accepted: 8 January 2018; Published: 10 January 2018

Abstract: In this work the high cycle fatigue behavior of a particulate reinforced 2124 aluminum alloy, manufactured by powder metallurgy, is investigated. SiC particles with a size of 3 μm and 300 nm and a volume fraction of 5 and 25 vol %, respectively, were used as reinforcement component. The present study is focused on the fatigue strength and the influence of particle size and temperature. Systematic work is done by comparing the unreinforced alloy and the reinforced conditions. All of the material conditions are characterized by electron microscopy and tensile and fatigue testing at room temperature and at 180 °C. With an increase in temperature the tensile and the fatigue strength decrease, regardless of particle size and volume fraction due to the lower matrix strength. The combination of 25 vol % SiC particle fraction with 3 μm size proved to be most suitable to achieve a major fatigue performance at room temperature and at 180 °C. The fatigue strength is increased by 40% when compared to the unreinforced alloy, as it is assumed the interparticle spacing for this condition reaches a critical value then.

Keywords: metal matrix composite; high cycle fatigue; high temperature properties; particulate reinforcement; aluminum alloy

---

1. Introduction

The performance requirements of materials for advanced engineering applications in the aerospace and automotive industry call for lightweight structural composite materials. Aluminum matrix composites (AMCs) are designed to meet these requirements such as a high specific strength and stiffness [1–3], an excellent fatigue performance [4–7], and an enhanced thermal stability [8–10].

The mechanical properties and the fatigue performance of the AMCs are strongly determined by different factors. Composites processed by powder metallurgy, as used in this study, exhibit a homogeneous particle distribution, and therefore superior mechanical properties and an outstanding fatigue performance when compared to other manufacturing processes [11]. Tensile strength and fatigue resistance of the composites are also strongly influenced by the aging condition and the matrix microstructure [12,13]. Particle shape [14,15], particle size, and volume fraction [16–20] are critical factors for the mechanical properties and fatigue behavior of the AMCs. Due to an enhanced load transfer from the softer matrix to the stiffer particles, an increase in particle volume fraction and a decrease in particle size lead to a significant increase in fatigue strength [4–7,21]. Further, particle size and interparticle spacing are determining factors for the fracture behavior [22,23]. At room temperatures, decohesion of the particles from the matrix is unlikely to occur, due to the high interface strength. Therefore, the probability of particle failure increases [21,24]. Larger particles provide a minor resistance against particle failure [4,15,25] and fatigue cracks initiate preferably at them due to the higher local stress concentration [12,17,26]. In contrast, for composites with smaller particles, the stress distribution is more homogeneous due to minor local stress concentrations and smaller interparticle spacing [27]. With increasing temperature, interface decohesion is generally more likely to occur, but larger particles are still prone to fracture [28–30].

As a result, to enhance the fatigue limit and minimize critical factors for crack initiation, the usage of small particles is required. Nonetheless, there is only limited data in literature on particle sizes smaller than 5 µm. Therefore, the purpose of the present study is to examine the influence of particles with 3 µm and 300 nm size with two different particle volume fractions. The fatigue tests were carried out at room temperature and 180 °C and the effects of temperature and particle size on the fatigue strength are discussed.

2. Materials and Methods

2.1. Material

In this study a 2124 aluminum alloy reinforced with two different particle volumes and particle sizes, respectively, was investigated. The material conditions were processed by high energy ball milling, hot isostatic pressing and forging by Materion Aerospace Composites AMC (Farnborough, UK) and were provided as plates with the dimensions given in Figure 1. The chemical composition of the matrix material is given in Table 1. SiC particles with 3 µm and 300 nm size were used as reinforcement. In this study, five material conditions were investigated: Unreinforced, reinforced with 5 vol % and 25 vol % 3 µm SiC particles and reinforced with 5 vol % and 25 vol % 300 nm SiC particles. All of the tested conditions were solid-solution treated at 505 °C for 60 min, and subsequently cold-aged at room temperature (RT) for 100 h.

Table 1. Chemical composition of the 2124 aluminum powder alloy.

Element	Al	Cu	Mg	Mn	Si	Fe	Cr	Ti	Zn	Others
wt %	91.25	4.9	1.8	0.9	0.2	0.3	0.1	0.15	0.25	0.15

2.2. Methods of Mechanical Testing and Electron Microscopy

Quasi-static tensile tests were performed in a Zwick-Roell servohydraulic testing machine (Zwick, Ulm, Germany) at a strain rate of $10^{-3} \cdot s^{-1}$ at room temperature and at 180 °C. For tensile testing cylindrical specimens were used with a cross section of 3.5 mm and a gauge length of 10.5 mm (sample orientation is given in Figure 1). For each condition, three samples were tested.

High cycle fatigue tests were performed on a RUMUL Testronic resonant testing machine (Russenberger Prüfmaschinen AG, Neuhausen am Rheinfall, Switzerland) under tension-tension loading with a load ratio of $R = 0.1$. The fatigue tests were carried out until the endurance limit of $N_D = 10^7$ cycles (approx. 27 h testing time) was reached or until a crack occurred, which was detected by a drop in the resonant frequency of 2 Hz or more. Axial fatigue specimens with 4.0 mm minimum diameter were used for fatigue testing (sample orientation and specimen geometry are given in Figure 1). For testing at 180 °C, the tensile and fatigue specimens were preheated for 30 min at this temperature. This led to a further aging of the formerly underaged matrix.

From all of the conditions, samples for microstructural analysis were extracted from the forged material, as shown in Figure 1. These samples were analyzed by quadrant back scatter diffraction (QBSD) at 20 kV using a Zeiss Neon 40 field emission microscope (Carl Zeiss MicroImaging GmbH, Jena, Germany).

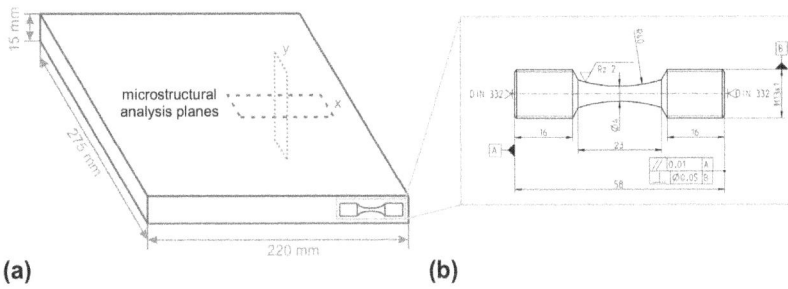

Figure 1. Schematic figure of the dimensions of the forged plate, the sample orientation of the axial specimens for tensile and fatigue testing and the investigated planes for the microstructural analysis (**a**) and the specimen geometry for fatigue testing (**b**).

3. Results and Discussion

3.1. Microstructure

All of the tested material conditions exhibit a homogeneous microstructure with numerous coarse Al_2Cu precipitates. The reinforced conditions show areas without reinforcement components with a width of 30–150 μm and a height of about 10–20 μm (see Figure 2). The particle distribution is rather homogeneous for all reinforced conditions. The SiC particles are irregularly shaped and exhibit an intact interface to the aluminum matrix (see Figure 3).

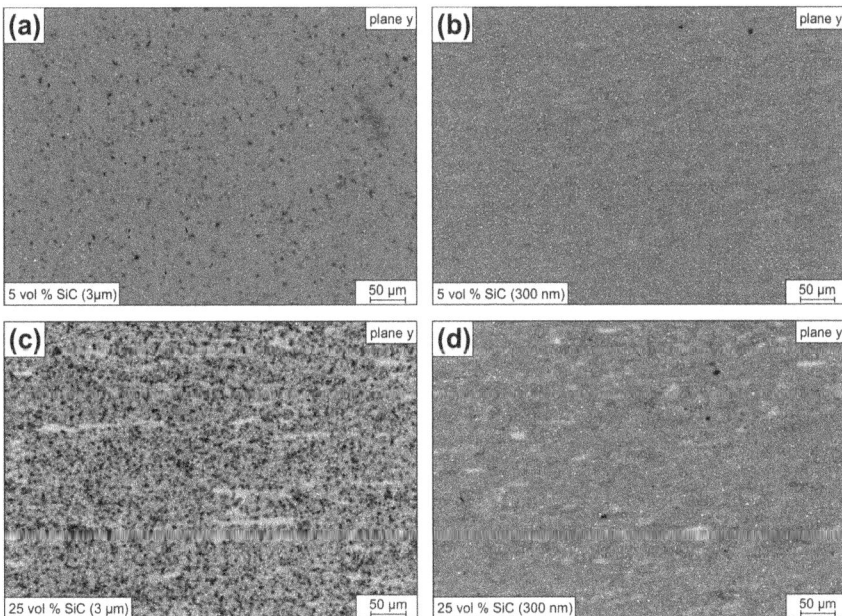

Figure 2. Quadrant back scatter diffraction (QBSD) micrographs of the 2124 aluminum alloy reinforced with (**a,b**) 5 vol % and (**c,d**) 25 vol % SiC particles with a size of (**a,c**) 3 μm and (**b,d**) 300 nm. The reinforced conditions exhibit areas without reinforcement component (matrix is light grey, particles are dark grey, Al_2Cu precipitates are white).

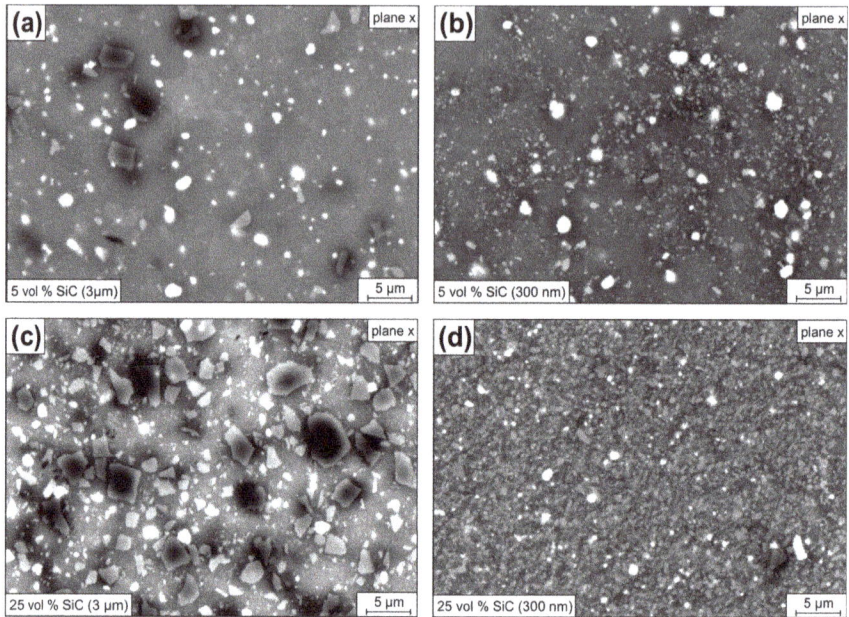

Figure 3. QBSD micrographs of the 2124 aluminum alloy reinforced with (**a**,**b**) 5 vol % and (**c**,**d**) 25 vol % SiC particles with a size of (**a**,**c**) 3 μm and (**b**,**d**) 300 nm. The SiC particles are irregularly shaped and finely dispersed: Matrix and particles exhibit an intact interface.

3.2. Tensile Testing

In Figure 4 and Table 2 the tensile properties for all of the tested material conditions (in underaged heat treatment condition) and temperatures are given. At room temperature, the unreinforced alloy exhibits the lowest yield strength with 341 MPa, as well as the highest uniform elongation with 17%, and therefore the highest ductility for the tested material conditions (see Figure 4a). The presence of a 5% volume fraction of reinforcement decreases the uniform elongation by approximately a third. For 25 vol % SiC particles, the uniform elongation is drastically reduced to only 1%. For 5 vol % particle fraction, a reduction in particle size from 3 μm to 300 nm increases the yield strength by 5% to 378 MPa. The strengthening effect caused by particle size reduction is more pronounced for 25 vol % particle fraction, as the yield strength for the material with 300 nm particles is increased by 38% to 633 MPa, which is the highest yield strength of all tested material conditions. This is an increase by 85% compared to the unreinforced alloy. Clearly, for the reinforced material the determining factor for the elongation is the reinforcement volume fraction, whereas both particle size and volume fraction determine the strength.

The increase in strength due to the particulate reinforcement is caused mainly by the load transfer from the matrix to the stiffer reinforcement component [7,23]. Further, dispersion strengthening occurs, which leads to two main mechanisms. Dislocation generation and the high dislocation density due to the difference in the thermal expansion coefficient of the matrix and the SiC particle [31], as well as the impediment of the dislocation movement due to the high dislocation density and the reinforcement component [4,12] act as strengthening mechanisms. An increase in particle volume fraction and a decrease in particle size lead to an enhancement of this effect, and therefore to a further increase in strength and a decrease in ductility [1,32]. The minor ductility of the reinforced material and the further decrease with an increase in particle volume fraction is caused by the major residual stresses induced by the reinforcement component [27].

At 180 °C all of the material conditions exhibit a decrease in strength and an increase in ductility when compared to the room temperature properties (see Figure 4b). The unreinforced alloy exhibits the lowest yield strength with 282 MPa and the highest uniform elongation with approximately 18%. At 180 °C the influence of particle size on the yield strength decreases. The reinforced material with 5 vol % particle fraction exhibit only an 18% higher yield strength if compared to the unreinforced alloy. Reinforcement with 25 vol % leads to the highest yield strength of about 420 MPa, which is twice as high as for the unreinforced alloy. At 180 °C, smaller particle sizes result in a higher ductility. The elongation to failure is at least one third higher in the material with smaller reinforcements, than in the material with 3 mm particles. Clearly, the uniform elongation for 5 vol % particle fraction and both particle sizes is nearly the same as at room temperature.

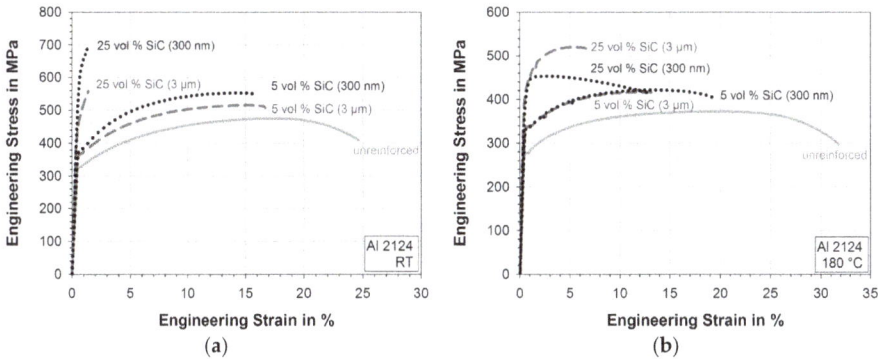

Figure 4. Tensile behavior of the unreinforced and reinforced 2124 aluminum alloy (**a**) at room temperature and (**b**) at 180 °C. Figure shows one representative curve for each condition.

Table 2. Mechanical properties determined by tensile testing of the unreinforced and reinforced 2124 aluminum alloy. The deviation is given in absolute values.

Condition of the 2124 Aluminum Alloy	Temperature	Yield Strength in MPa	Ultimate Tensile Strength in MPa	Uniform Elongation in %	Elongation to Failure in %
unreinforced	RT	341 ± 35	471 ± 8	16.7 ± 1.0	24.8 ± 1.1
5 vol % SiC (3 µm)	RT	358 ± 8	508 ± 9	14.0 ± 1.0	15.6 ± 0.8
25 vol % SiC (3 µm)	RT	460 ± 6	573 ± 62	1.5 ± 1.2	1.5 ± 1.2
5 vol % SiC (300 nm)	RT	378 ± 1	552 ± 1	13.3 ± 1	14.5 ± 1.0
25 vol % SiC (300 nm)	RT	633 ± 10	727 ± 33	1.0 ± 0.4	1.0 ± 0.4
unreinforced	180 °C	282 ± 15	380 ± 17	17.4 ± 1.6	28.1 ± 3.1
5 vol % SiC (3 µm)	180 °C	332 ± 4	413 ± 4	11.9 ± 0.5	13.5 ± 1.0
25 vol % SiC (3 µm)	180 °C	426 ± 6	529 ± 11	1.7 ± 0.1	6.0 ± 1.6
5 vol % SiC (300 nm)	180 °C	337 ± 5	423 ± 5	12.5 ± 0.3	18.8 ± 1.6
25 vol % SiC (300 nm)	180 °C	419 ± 1	458 ± 5	2.0 ± 0.3	13.1 ± 1.2

With an increasing temperature, the matrix properties become of more influence for the tensile properties and the strength decreases, regardless of the particle volume fraction and the particle size [33]. The influence of particle size on the strength decreases due to the softening of the overaged matrix and the increase in relaxation of residual stresses and local stress concentration [30]. The probability for particle fracture decreases and matrix failure near the interface or interfacial decohesion becomes more prominent [29,34]. The ductility for reinforced material with smaller particles is increased due to their smaller interparticle spacing, and therefore a more homogeneous distribution of the plastic strain in the matrix [27,30].

3.3. High Cycle Fatigue Behavior

The high cycle fatigue behavior of the tested material conditions and its dependence on temperature is shown in Figure 5 and the fatigue strength for $N_D = 10^7$ cycles is listed in Table 3.

At room temperature, the unreinforced alloy exhibits a fatigue strength of 280 MPa (see Figure 5a). The fatigue strength of both 5 vol % reinforced conditions is smaller than for the unreinforced alloy. The condition with 3 μm particle size exhibits the lowest fatigue strength with 250 MPa. Reinforcement with 25 vol % 3 μm SiC particles increases the fatigue strength by 40% if compared to the unreinforced alloy. Reducing the particle size to 300 nm leads to an additional increase by 6%, and therefore to the highest fatigue strength of all conditions with 410 MPa.

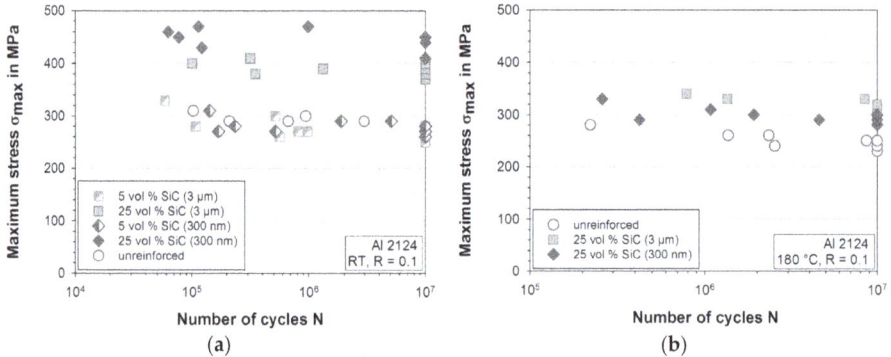

Figure 5. High cycle fatigue behavior of the unreinforced and reinforced 2124 aluminum alloy at $R = 0.1$ and (**a**) at room temperature and (**b**) at 180 °C.

Table 3. Fatigue limit at $N_D = 10^7$ cycles of the unreinforced and reinforced 2124 aluminum alloy at RT and load ratio $R = 0.1$.

Condition of the 2124 Aluminum Alloy	Temperature	Maximum Stress σ_{max} in MPa	Reduction in Fatigue Limit in % [1]	Increase in Fatigue Limit in % [2]
Unreinforced	RT	280	-	-
5 vol % SiC (3 μm)	RT	250	10.7	-
25 vol % SiC (3 μm)	RT	390	-	39.3
5 vol % SiC (300 nm)	RT	270	3.6	-
25 vol % SiC (300 nm)	RT	410	-	46.4
Unreinforced	180 °C	230	17.9	-
25 vol % SiC (3 μm)	180 °C	320	-	39.1
25 vol % SiC (300 nm)	180 °C	280	-	21.7

[1] referring to the unreinforced alloy at room temperature. [2] referring to the unreinforced alloy at the respective temperature.

A high yield and tensile strength of the particle reinforced material do not necessarily result in a high fatigue strength [6]. A deteriorated fatigue performance of the composites in comparison to the unreinforced alloy can be explained by a high defect density, particle clustering, or a high porosity of the material [7,35]. As these effects were not noticeable for the investigated conditions, the minor fatigue strength of the reinforced conditions with 5 vol % particle fraction is attributed to the local stress raisers induced by the reinforcement component. This diminishes the beneficial strengthening effect of the reinforcement. The minimal fatigue strength of the reinforced condition with 5 vol % and 3 μm particles can be explained by the further increase of the local stress intensity at the particle-matrix-interface with an increase in particle size [26]. Also, the particle shape essentially affects the stress concentration [14]. Sharp edges of the irregularly shaped particles and slight imperfections of the particle surface cause a significant increase in the internal stresses. Therefore, an early formation of initial cracks can further be a reason for the minor fatigue strength of the reinforced conditions with 5 vol % particle fraction [21]. The increase in fatigue strength with an increase in particle volume content and a decrease in particle size is a well known effect [4–7,21]. An increase in volume fraction enables a higher load transfer from the matrix to the stiffer reinforcement component [6,7]. An additional decrease in particle size leads to a decrease in the interparticle spacing, which prevents the formation of reversible slip bands when a critical value is reached [5].

In Figure 5b, the high cycle fatigue behavior at 180 °C is shown. The reinforced conditions with 5 vol % SiC particles were not tested at this temperature due to their minor fatigue performance at room temperature. Due to the long testing time, overaging processes occur as observed for the tensile tests. It is assumed, that all of the tested material conditions are equally overaged after preheating for 30 min. An increase in temperature leads for the unreinforced alloy and the reinforced material to a decrease in fatigue strength. The fatigue strength of the unreinforced alloy is 230 MPa, and therefore 18% lower than at room temperature. In contrast to fatigue at room temperature, the condition with 25 vol % and 3 µm particle size exhibits the highest fatigue strength with 320 MPa, which is an increase by 40% if compared to the unreinforced alloy at 180 °C. Reinforcement with 300 nm particles leads to a much smaller increase in fatigue strength by 22%.

At higher temperatures the stress concentrations and the influence of processing defects decreases and the matrix becomes the determining factor for the fatigue strength [7,33]. The lower matrix strength, due to the overaging during testing, causes a general reduction in fatigue strength for all of the tested conditions at 180 °C [12]. Fatigue crack initiation in the matrix is the primary failure mechanism [7,20,23,36]. With increasing temperature, decohesion between the particle and the matrix is enhanced and cracks are easily initiated [28,37]. Additionally, cracks are also generated by cyclic slip deformation for reinforced conditions with a particle size smaller 20 µm [33,36]. The crack initially propagates along a slip band and ahead of the crack tip microcracking and void formation in the matrix occur [37]. These microcracks join the main crack by matrix microvoid coalescence [23,37]. In addition, the plastic zone ahead of the fatigue crack front is larger than the average interparticle spacing and determines the fracture mechanisms [37]. The findings of [33], which state a declining influence of the particle volume fraction and of the particle size on the fatigue strength with increasing temperature, could not be fully confirmed by our work. Reinforcement with 300 nm particles led to a minor increase in fatigue strength in comparison to room temperature. This effect can be explained by the relieved dislocation cross slip motion with an increased temperature due to the small interparticle spacing. Additionally, the small interparticle spacing and the high particle volume fraction lead to a major amount of particles incorporated in the plastic zone ahead of the crack tip, and therefore to major void formation. In contrast, the interparticle spacing between the 3 µm particles is large enough to impede the dislocation movement and still small enough to limit the void growth [23]. It is suggested that 3 µm is an optimal value for the particle size at the given particle fraction of 25 vol % to maximize the matrix and fatigue strength. This explains the 40% higher fatigue strength of this condition if compared to the unreinforced alloy at room temperature and 180 °C. For the commonly used minimal particle size of 5 µm, the interparticle spacing is already large enough to enable void growth and dislocation slip motion between the particles. Therefore, the strengthening effect due to particle reinforcement is smaller at higher temperatures.

4. Conclusions

The influence of temperature and particle size on the high cycle fatigue behavior of the particulate reinforced 2124 aluminum alloy is investigated. SiC particles with 3 µm and 300 nm size and a volume fraction of 5 and 25 vol %, respectively, were used as reinforcement component. The tensile properties and the fatigue behavior of the unreinforced alloy and the four reinforced conditions were compared at room temperature and at 180 °C. Conclusions can be drawn as follows:

1. Generally, the tensile and the fatigue strength of the unreinforced and the reinforced material decrease with an increase in temperature. This is attributed to the increasing influence of the lower matrix strength, regardless of the particle volume fraction and the particle size.

2. Particulate reinforcement leads to an increase in tensile strength and a loss in ductility. A high particle volume fraction enhances this effect. At room temperature, a decrease in particle size leads to a further increase in tensile strength, whereas at 180 °C, the tensile strength is not affected by a decreased particle size, nevertheless, the ductility increases.

3. The room temperature fatigue strength of the reinforced conditions with 5 vol % SiC in 3 μm and 300 nm size was minor in comparison to the unreinforced alloy. Supposedly the local stress concentrations induced by the reinforcement component lead to an early formation of initial cracks.

4. The beneficial effect of an increased particle volume fraction and a decreased particle size on the fatigue strength could be confirmed for room temperature. The condition with 25 vol % particle fraction and 300 nm size exhibited the highest fatigue strength. Whereas, at 180 °C, reinforcement with 25 vol % SiC particles leads also to a significant increase in fatigue strength when compared to the unreinforced alloy, but the percentage increase was minor for the 300 nm particles if compared to the 3 μm particles.

5. The combination of 25 vol % SiC particle fraction with 3 μm size proved to be most suitable for a major fatigue performance at room temperature and at 180 °C. It is assumed, that the interparticle spacing for this combination of particle size and fraction is large enough to impede dislocation movement and still small enough to limit void formation. Therefore, the fatigue strength was improved by 40% in comparison to the unreinforced alloy at both testing temperatures.

Acknowledgments: The authors gratefully acknowledge funding of the Collaborative Research Center SFB 692 received from the German Research Foundation (Deutsche Forschungsgemeinschaft DFG).

Author Contributions: Lisa Winter designed, performed and analyzed the experiments and is the primary author of the paper. Kristin Hockauf discussed the results and analysis with the author. Thomas Lampke supervised the work.

Conflicts of Interest: The authors declare no conflict of interest.

References

1. Doel, T.J.A.; Bowen, P. Tensile properties of particulate-reinforced metal matrix composites. *Compos. Part A Appl. Sci. Manuf.* **1996**, *27*, 655–665. [CrossRef]
2. Ceschini, L.; Morri, A.; Cocomazzi, R.; Troiani, E. Room and high temperature tensile tests on the AA6061/10vol.%Al$_2$O$_3$p and AA7005/20vol.%Al$_2$O$_3$p composites. *Mater. Sci. Eng. Technol.* **2003**, *34*, 370–374. [CrossRef]
3. Ceschini, L.; Minak, G.; Morri, A. Tensile and fatigue properties of the AA6061/20vol.% Al$_2$O$_3$p and AA7005/10vol.% Al$_2$O$_3$p composites. *Compos. Sci. Technol.* **2006**, *66*, 333–342. [CrossRef]
4. Hall, J.N.; Jones, J.W.; Sachdev, A.K. Particle size, volume fraction and matrix strength effects on fatigue behavior and particle fracture in 2124 aluminum-SiCp composites. *Mater. Sci. Eng. A* **1994**, *183*, 69–80. [CrossRef]
5. Chawla, N.; Andres, C.; Jones, J.W.; Allison, J.E. Effect of SiC volume fraction and particle size on the fatigue resistance of a 2080 Al/SiC composite. *Metall. Mater. Trans. A* **1998**, *29*, 2843–2854. [CrossRef]
6. Chawla, N.; Allison, J.E. Fatigue of Particle Reinforced Materials. In *Encyclopedia of Materials: Science and Technology*, 2nd ed.; Elsevier: Amsterdam, The Netherlands, 2001; pp. 2967–2971.
7. Llorca, J. Fatigue of particle-and whisker-reinforced metal-matrix composites. *Prog. Mater. Sci.* **2002**, *47*, 283–353. [CrossRef]
8. Furukawa, M.; Wang, J.; Horita, Z.; Nemoto, M.; Ma, Y.; Langdon, T.G. An investigation of strain hardening and creep in an Al-6061/Al$_2$O$_3$ metal matrix composite. *Metall. Mater. Trans. A* **1995**, *26*, 633–639. [CrossRef]
9. Li, Y.; Langdon, T.G. Creep behavior of an Al-6061 metal matrix composite reinforced with alumina particulates. *Acta Mater.* **1997**, *45*, 4797–4806. [CrossRef]
10. Tjong, S.C.; Ma, Z.Y. The high-temperature creep behaviour of aluminium-matrix composites reinforced with SiC, Al$_2$O$_3$ and TiB$_2$ particles. *Compos. Sci. Technol.* **1997**, *57*, 697–702. [CrossRef]

11. Park, B.G.; Crosky, A.G.; Hellier, A.K. High cycle fatigue behaviour of microsphere Al_2O_3–Al particulate metal matrix composites. *Compos. Part B Eng.* **2008**, *39*, 1257–1269. [CrossRef]
12. Chawla, N.; Habel, U.; Shen, Y.-L.; Andres, C.; Jones, J.W.; Allison, J.E. The Effect of Matrix Microstructure on the Tensile and Fatigue Behavior of SiC Particle-Reinforced 2080 AI Matrix Composites. *Metall. Mater. Trans. A* **2000**, *31*, 531–540. [CrossRef]
13. Srivatsan, T.S.; Mattingly, J. Influence of heat treatment on the tensile properties and fracture behaviour of an aluminium alloy-ceramic particle composite. *J. Mater. Sci.* **1993**, *28*, 611–620. [CrossRef]
14. Romanova, V.A.; Balokhonov, R.R.; Schmauder, S. The influence of the reinforcing particle shape and interface strength on the fracture behavior of a metal matrix composite. *Acta Mater.* **2009**, *57*, 97–107. [CrossRef]
15. Zhang, P.; Li, F. Effect of particle characteristics on deformation of particle reinforced metal matrix composites. *Trans. Nonferrous Met. Soc. China* **2010**, *20*, 655–661. [CrossRef]
16. Xue, Z.; Huang, Y.; Li, M. Particle size effect in metallic materials: A study by the theory of mechanism-based strain gradient plasticity. *Acta Mater.* **2002**, *50*, 149–160. [CrossRef]
17. Huang, M.; Li, Z. Size effects on stress concentration induced by a prolate ellipsoidal particle and void nucleation mechanism. *Int. J. Plast.* **2005**, *21*, 1568–1590. [CrossRef]
18. Köhler, L.; Hockauf, K.; Lampke, T. Influence of Particulate Reinforcement and Equal-Channel Angular Pressing on Fatigue Crack Growth of an Aluminum Alloy. *Metals (Basel)* **2015**, *5*, 790–801. [CrossRef]
19. Shyong, J.H.; Derby, B. The deformation characteristics of SiC particulate-reinforced aluminium alloy 6061. *Mater. Sci. Eng. A* **1995**, *197*, 11–18. [CrossRef]
20. Shin, C.S.; Huang, J.C. Effect of temper, specimen orientation and test temperature on the tensile and fatigue properties of SiC particles reinforced PM 6061 Al alloy. *Int. J. Fatigue* **2010**, *32*, 1573–1581. [CrossRef]
21. Papakyriacou, M.; Mayer, H.; Stanzl-Tschegg, S.; Groschl, M. Fatigue properties of Al_2O_3-particle-reinforced 6061 aluminium alloy in the high-cycle regime. *Int. J. Fatigue* **1996**, *18*, 475–481. [CrossRef]
22. Kamat, S.V.; Hirth, J.P.; Mehrabian, R. Mechanical properties of particulate-reinforced aluminum-matrix composites. *Acta Metall.* **1989**, *37*, 2395–2402. [CrossRef]
23. Milan, M.T.; Bowen, P. Tensile and Fracture Toughness Properties of SiC_p Reinforced Al Alloys: Effects of Particle Size, Particle Volume Fraction, and Matrix Strength. *J. Mater. Eng. Perform.* **2004**, *13*, 775–783. [CrossRef]
24. Flom, Y.; Arsenault, R.J. Interfacial bond strength in an aluminium alloy 6061-SiC composite. *Mater. Sci. Eng.* **1986**, *77*, 191–197. [CrossRef]
25. Mummery, P.; Derby, B. The influence of microstructure on the fracture behaviour of particulate metal matrix composites. *Mater. Sci. Eng. A* **1991**, *135*, 221–224. [CrossRef]
26. Tokaji, K.; Shiota, H.; Kobayashi, K. Effect of particle size on fatigue behaviour in SiC particulate-reinforced aluminium alloy composites. *Fatigue Fract. Eng. Mater. Struct.* **1999**, *22*, 281–288. [CrossRef]
27. Bouafia, F.; Serier, B.; Bouiadjra, B.A.B. Finite element analysis of the thermal residual stresses of SiC particle reinforced aluminum composite. *Comput. Mater. Sci.* **2012**, *54*, 195–203. [CrossRef]
28. Hadianfard, M.J.; Healy, J.; Mai, Y.-W. Temperature effect on fracture behaviour of an alumina particulate-reinforced 6061-aluminium composite. *Appl. Compos. Mater.* **1994**, *1*, 93–113. [CrossRef]
29. Poza, P.; Llorca, J. Fracture toughness and fracture mechanisms of Al Al_2O_3 composites at cryogenic and elevated temperatures. *Mater. Sci. Eng. A* **1996**, *206*, 183–193. [CrossRef]
30. Han, N.L.; Wang, Z.G.; Zhang, G.D. Effect of reinforcement size on the elevated-temperature tensile properties and low-cycle fatigue behavior of particulate SiC/Al composites. *Compos. Sci. Technol.* **1997**, *57*, 1491–1499. [CrossRef]
31. Vogelsang, M.; Aresenault, R.J.; Fisher, R.M. An In Situ HVEM Study of Dislocation Generation at Al/SiC Interfaces in Metal Matrix Composites. *Metall. Trans. A* **1986**, *17A*, 379–389. [CrossRef]
32. Knowles, A.J.; Jiang, X.; Galano, M.; Audebert, F. Microstructure and mechanical properties of 6061 Al alloy based composites with SiC nanoparticles. *J. Alloys Compd.* **2014**, *615*, 401–405. [CrossRef]
33. Uematsu, Y.; Tokaji, K.; Kawamura, M. Fatigue behaviour of SiC-particulate-reinforced aluminium alloy composites with different particle sizes at elevated temperatures. *Compos. Sci. Technol.* **2008**, *68*, 2785–2791. [CrossRef]
34. Biermann, H.; Kemnitzer, M.; Hartmann, O. On the temperature dependence of the fatigue and damage behaviour of a particulate-reinforced metal-matrix composite. *Mater. Sci. Eng. A* **2001**, *319–321*, 671–674. [CrossRef]

35. Vyletel, G.M.; Allison, J.E.; Van, D.C.A. The effect of matrix microstructure on cyclic response and fatigue behavior of particle—Reinforced 2219 aluminum: Part I. room temperature behavior. *Metall. Mater. Trans. A* **1995**, *26*, 3143–3154. [CrossRef]
36. Nieh, T.G.; Lesuer, D.R.; Syn, C.K. Tensile and Fatigue Properties of a 25 vol% SiC Particulate Reinforced 6090 Al Composite at 300 °C. *Scr. Metall. Mater.* **1995**, *32*, 707–712. [CrossRef]
37. Li, C.; Ellyin, F. Fatigue damage and its localization in particulate metal matrix composites. *Mater. Sci. Eng. A* **1996**, *214*, 115–121. [CrossRef]

metals

MDPI

Article
On the PLC Effect in a Particle Reinforced AA2017 Alloy

Markus Härtel *, Christian Illgen, Philipp Frint and Martin Franz-Xaver Wagner

Institute of Materials Science and Engineering, Chemnitz University of Technology, D-09107 Chemnitz, Germany; christian.illgen@mb.tu-chemnitz.de (C.I.); philipp.frint@mb.tu-chemnitz.de (P.F.); martin.wagner@mb.tu-chemnitz.de (M.F.-X.W.)
* Correspondence: markus.haertel@mb.tu-chemnitz.de; Tel.: +49-371-531-32532

Received: 15 December 2017; Accepted: 22 January 2018; Published: 25 January 2018

Abstract: The Portevin–Le Châtelier (PLC) effect often results in serrated plastic flow during tensile testing of aluminum alloys. Its magnitude and characteristics are often sensitive to a material's heat treatment condition and to the applied strain rate and deformation temperature. In this study, we analyze the plastic deformation behavior of an age-hardenable Al-Cu alloy (AA2017) and of a particle reinforced AA2017 alloy (10 vol. % SiC) in two different conditions: solid solution annealed (W) and naturally aged (T4). For the W-condition of both materials, pronounced serrated flow is observed, while both T4-conditions do not show distinct serrations. It is also found that a reduction of the testing temperature (-60 °C, -196 °C) shifts the onset of serrations to larger plastic strains and additionally reduces their amplitude. Furthermore, compressive jump tests (with alternating strain rates) at room temperature confirm a negative strain rate sensitivity for the W-condition. The occurring PLC effect, as well as the propagation of the corresponding PLC bands in the W-condition, is finally characterized by digital image correlation (DIC) and by acoustic emission measurements during tensile testing. The formation of PLC bands in the reinforced material is accompanied by distinct stress drops as well as by perceptible acoustic emission, and the experimental results clearly show that only type A PLC bands occur during testing at room temperature (RT).

Keywords: serrated flow; PLC effect; dynamic strain aging; particle reinforcement; acoustic emission; jump tests; aluminum alloy

---

1. Introduction

Plastic instabilities, which occur in many age-hardenable aluminum alloys [1–13] within a certain regime of strain rates and testing temperatures, are often associated with the observation of repeated stress serrations in the stress–strain curves of tensile or compression tests. The so-called "jerky" or "serrated" flow is one of the most distinctive examples of plastic instability in dilute alloys. It is commonly rationalized by the dynamic interactions between solute atoms and mobile dislocations, i.e., dynamic strain aging processes [1,3–6,9,11,12,14–23]. Early observations of this phenomenon trace back to Le Châtelier who observed stress serration in the plastic flow of mild steel at elevated temperatures [24]. First studies of serrated flow in aluminum alloys were performed by Portevin and Le Châtelier [25] at ambient temperature but under different strain rates. Today, an occurrence of stress serrations in the plastic flow behavior as a result of dynamic strain aging processes is, therefore, usually referred to as the Portevin–Le Châtelier effect (PLC effect).

The PLC effect often leads to the nucleation and propagation of localized deformation bands on a macroscopic scale. Localization occurs if the local strain rate of the material exceeds the macroscopically applied strain rates during deformation [26]. A material's susceptibility for localization is closely related to its strain rate sensitivity m: Negative strain rate sensitivities of the flow stress, as well as a cooperative dislocation motion, are necessary requirements for the existence of a PLC effect [27–30]. Negative strain rate sensitivities of the material typically are the result of dynamic interactions between gliding dislocations and mobile solute atoms. Their interaction leads to repeating pinning and unpinning processes [14,15,31]. Therefore, local variations of mobile dislocation density increase the (negative) strain rate sensitivity. However, many other parameters can also influence the magnitude of an actual PLC effect: the chemical composition, the amount and type of solute atoms, texture, grain size, strain hardening rate, heat treatment condition, and especially strain rate as well as temperature [3,16,18,21,32]. Serrated plastic flow may also influence formability: PLC effects in sheet metal forming lead, e.g., to an optical degradation (strain marks). In extrusion, PLC effects lead to a reduced formability, but with thermal suppression of those deformation bands, those negative aspects of localized plastic flow can be reduced [10,32].

In this paper, we study the influence of different heat treatments, deformation temperatures and applied strain rates on the magnitude of a potential PLC effect in a high-strength, age-hardenable Al-Cu alloy (AA2017), and in a particle reinforced AA2017 alloy (10 vol. % SiC) in two different conditions. Studies on the general mechanical properties, the deformation behavior and the accompanying microstructural aspects have been the focus of recent research efforts [10,33,34]. The influence of a finely dispersed and homogeneously distributed particle reinforcement on the occurrence and magnitude of PLC effects, however, has not been studied in detail so far. The present study documents in detail that a PLC effect only occurs in solution annealed conditions, and that a reduced testing temperature leads to a suppression of PLC band formation.

2. Materials and Methods

The mechanical behavior of two different age-hardenable alloys was studied in this work. The chemical compositions (wt. %) of the examined AA2017 alloy as well as of the particle reinforced AA2017 alloy (with 10 vol. % SiC; particle size less than 2 µm) are given in Table 1. Both materials were obtained as commercial, gas-atomized, spherical powders with a particle size below 100 µm [35]. The powders were milled for about 4 h in a high energy ball mill (Simoloyer1 CM08 by Zoz GmbH, Wenden, Germany), followed by hot isostatic pressing (HIP) at 450 °C for 3 h and at a pressure of 1100 bar [35]. Semi-finished products were then generated by extrusion of the mixed powders to billets with a square cross-section of 15 × 15 mm². Further details on the fabrication procedures and particularly on the morphology and on the homogeneous spatial distribution of the reinforcing particles are given in [35]. For characterization of the thermo-mechanical behavior and of the strain rate sensitivity, cylindrical tensile samples (with a gauge length of 10.5 mm and a diameter of 3.5 mm) and cylindrical compression specimens (with an aspect ratio of one, diameter 5 mm) were machined from the billets parallel to the extrusion direction.

Table 1. Chemical composition of the age-hardenable Al-Cu alloy AA2017 and of the particle reinforced AA2017 composite material (with 10 vol. % SiC). The base compositions of both materials (i.e., of the alloy vs. the matrix-forming alloy of the composite) are very similar, allowing for a direct comparison of the mechanical behavior.

Element	Cu	Mg	Mn	Fe	Si	Cr	Ni	Ti	Zn	Al
AA2017 Content in wt. %	3.83	0.71	0.64	0.17	0.065	0.017	0.0049	0.0025	<0.00100	bal.
AA2017 (10 vol. % SiC) Content in wt. %	3.81	0.68	0.49	0.27	8.26	0.010	0.016	0.016	0.11	

As observed in previous studies [6,16,32,36], different heat treatment conditions can strongly affect the occurrence and magnitude of PLC effects. In order to provide a well-defined heat treatment condition, the materials were heat treated to generate two conditions: solid solution annealed (W) and naturally aged (T4). For the W-condition, the material was solid solution annealed for 120 min at 505 °C and quenched in ice water. To suppress further natural aging processes prior to mechanical testing, all specimens were stored in liquid nitrogen (at −196 °C) after the heat treatment. The naturally aged condition (T4) was solution annealed with the same parameters and subsequently naturally aged at room temperature (RT) for two weeks [35].

To characterize the deformation behavior of both materials, we performed strain-controlled uniaxial tensile tests (Zwick/Roell 20 kN tensile testing machine, Ulm, Germany) for both conditions W and T4. For all tensile tests, the temperature was RT and the strain rate was 10^{-3} s^{-1}. Besides the heat treatment condition, the applied strain rate governs a potential PLC effect. Particularly a negative strain rate sensitivity increases the potential of the occurrence of PLC effects [1,2,4,5,15,32,37,38]. In order to characterize the susceptibility of both materials to strain localization, we therefore also performed strain-controlled uniaxial compression tests (Zwick/Roell 20 kN tensile testing machine, Ulm, Germany) at RT with a sudden change of strain rate (jump tests). During those jump tests, the strain rate was repeatedly changed by one order of magnitude. The applied strain rates were in a range from 10^{-5} s^{-1} to 10^{-2} s^{-1} for the both investigated material conditions. The strain rate sensitivity value m was then determined from the true stress-strain curves by evaluating stress increments at the points of strain rate changes:

$$m = \frac{\Delta \ln \sigma}{\Delta \ln \dot{\varepsilon}} = \frac{\ln \sigma_2 - \ln \sigma_1}{\ln \dot{\varepsilon}_2 - \ln \dot{\varepsilon}_1}. \tag{1}$$

In Equation (1), $\dot{\varepsilon}_1$ is representing the strain rate immediately prior to, and $\dot{\varepsilon}_2$ the strain rate immediately after, the change of strain rate. The values for σ_1 and σ_2 are the corresponding true stresses approximated by the application of tangents to the regions of constant strain rate.

Since testing temperature has a considerable influence on the magnitude of serrated flow [32], the uniaxial tensile tests were performed at different temperatures (RT, -60 °C, and -196 °C) at a constant strain rate of 10^{-3} s^{-1}. A special double–ring cooling device placed around the tensile specimen was used to adjust testing temperatures below RT, Figure 1. For testing temperatures of -60 °C, the inner ring of the device was filled with n-pentane (C_5H_{12}) until the specimen was completely immersed, Figure 1b. Cooling was performed by adding liquid nitrogen into the outer ring of the device. For the testing temperature of -196 °C, both chambers of the cooling device were filled with liquid nitrogen. Using a thermocouple (type K), the testing temperatures were controlled in situ. The thermocouple was located directly on the surface of the specimens. Strain values during these tests were determined from the crosshead displacement data of the tensile testing machine.

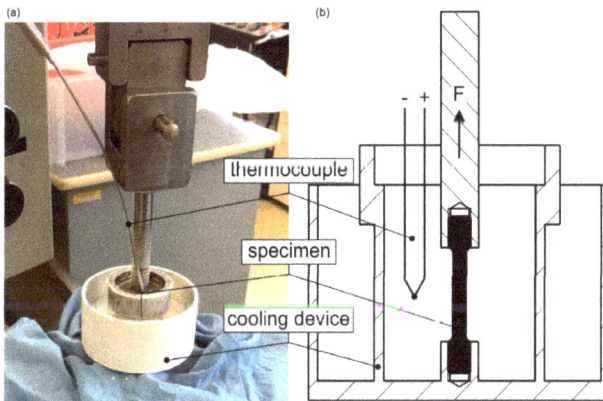

Figure 1. Experimental setup applied for tensile tests at different temperatures below room temperature (RT) as (**a**) photograph, and (**b**) as schematic drawing. The setup consists of a special double–ring cooling device surrounding the tensile specimen. The testing temperature is measured with a thermocouple placed on the sample surface.

Localized deformation can be documented by diffraction methods, 2D or 3D local strain mapping as well as by infrared thermal imaging methods [39–44]. In the present study, the local deformation behavior of the investigated material conditions was characterized in greater detail by additional tensile tests in combination with digital image correlation (DIC). DIC allows the documentation and analysis of the nucleation as well as the propagation of macroscopic deformation bands resulting from PLC effects by recording surface displacement fields and the corresponding strain maps. A fine speckle pattern for image analysis was produced on the sample surfaces using a graphite spray (a close-up view of the sample surface is shown in the video provided as supplementary material Video S1: Video S1: PLC effect reinforced AA2017 using DIC and AE). Finally, a highly sensitive unidirectional stereo microphone (XYH-6 X/Y Capsule by Zoom™, Hauppauge, NY, USA, 24 bit, up to 96 kHz) was used in the same setup (Figure 2) to measure acoustic emission signals (as integral acoustic intensity) based on earlier reports on acoustic emission related to PLC effects [2,3,15,18,45–47].

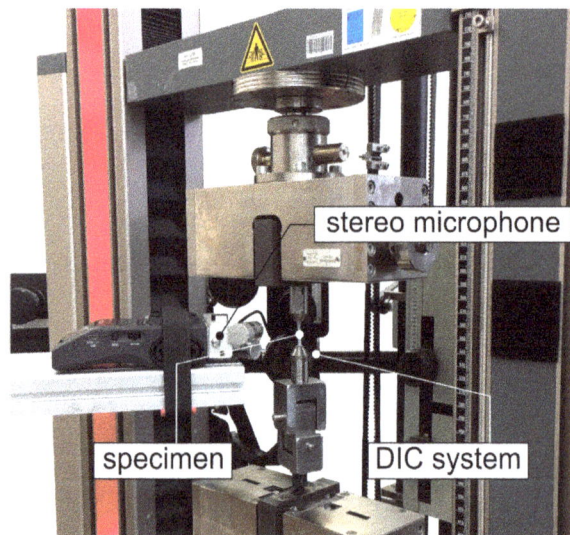

Figure 2. Experimental setup for digital image correlation (DIC) observations used during tensile testing. In addition to the optical measurement components, a highly sensitive unidirectional stereo microphone was used to measure the acoustic emission associated with The Portevin–Le Châtelier (PLC) effects.

3. Results and Discussion

Figure 3 shows tensile (engineering) stress-strain curves of both materials in the W- and T4-conditions. The data show that, as expected, both particle reinforcements and aging increase the strength of the material. In contrast to the T4-condition, the W-condition exhibits the typical serrated flow (PLC effect) for both materials. PLC effects are strongly influenced by the concentration of solute atoms (chemical composition and aging condition). Aging of aluminum alloys leads to precipitation, which considerably reduces the number of solute atoms. As a consequence, the formation of PLC bands is suppressed or delayed towards higher strains for the naturally aged condition T4 in Figure 3. This result is in good agreement with the results found in [3,6,16,32,48]. Another important finding is the different onset strain of the serrated flow for both materials: The plastic strain associated with the formation of the first PLC band is 6.5% for the AA2017 base material, but only 1.9% for the reinforced material. The onset of the PLC effect in the stress-strain curve is typically attributed to a necessary density of defects (basically dislocations and vacancies) that is required for dynamic strain aging [3,6,7,49–53]. Hence, an initial concentration of defects introduced by either quenching or by early plastic deformation is required to increase diffusion rates of the solute atoms and thus to trigger dislocation pinning. As discussed in [14,48], an earlier occurrence of stress serrations in the particle reinforced material (Figure 3) is in good agreement with this interpretation: The particle/matrix interfaces and the stress fields surrounding the particles locally lead to a rapid increase of the dislocation density [33] during plastic deformation. This increased number of defects allows for faster pipe diffusion processes [54–57] and—for the material studied here—accelerates dynamic strain aging at lower strain values. The extent of this effect likely depends on the volume fraction and the spatial arrangement of the dispersed particles [14,48].

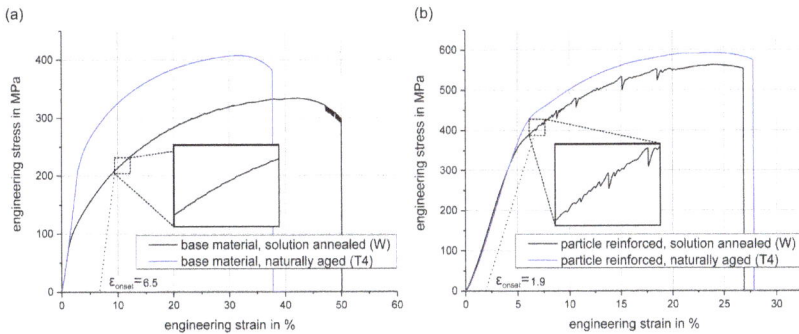

Figure 3. Stress-strain curves of (**a**) the AA2017 base material in the solution annealed condition W and in the naturally aged condition T4, and (**b**) of the particle reinforced AA2017 material in solution annealed condition W and in naturally aged condition T4. Only the W-conditions exhibit serrated flow. In the inset with a higher magnification, type A PLC serrations can be observed.

It is clear from Figure 3 that the characteristics of the serrations (i.e., the "shapes" of peaks and drops in the stress-strain curves) caused by localized deformation clearly differ in the two materials. Based on the characteristics of serrations, different types of PLC bands can be classified; the most common terminology distinguishes three different types [14,22,28,58,59]. Type A is characterized by repeated nucleation of single bands that move continuously from one side of the gauge length through the entire specimen. The corresponding stress-strain curves typically show a sudden increase, followed by a drop below the original stress level. Type B PLC bands are quickly arrested after nucleation; instead of widespread propagation, new bands are formed in the direct vicinity of the arrested bands. The stress-strain signal shows rapid serrations that oscillate around the original stress level. Type C PLC bands are associated with nucleation of randomly located bands within the gauge length and show repeated drops below the stress level followed by non-linear stress increases. For both AA2017 and AA2017 (10 vol. % SiC), we observed type A PLC bands in W-condition, see the higher magnification inset of Figure 3. The amplitudes of the serrations in Figure 3a (base material) are not as pronounced as in the inset in Figure 3b (particle reinforced material). This may be attributed to the stress fields of the particles, which represent obstacles for dislocation motion and thus increase the pinning effect. Between each large stress drop and the next pronounced serration in Figure 3b, we also observe additional serrated flow with lower amplitudes. These second-order serrations might, in theory, indicate the additional formation of type B PLC bands. However, such bands could not be recorded by our DIC observations. Therefore, we conclude that the propagation of the observed PLC bands (type A) is stopped temporarily by the stress fields of individual or clusters of several. This leads to additional pinning on the macro-scale (i.e., pinning of the propagating band as opposed to pinning of individual dislocations). We note that the amplitude of stress-strain serrations may also be influenced by the stiffness of the testing device, by the geometry of the specimen and by the surface quality of the specimen [8,17,21,45].

Figure 4 shows the results of the (compressive) jump tests of both materials in the solution annealed condition (W). The green and blue curves illustrate the true stress-true strain data (which are required for determining m-values) of the base material and of the particle reinforced material, respectively. In a range from 10^{-5} s^{-1} to 10^{-2} s^{-1} both materials exhibit a positive stress overshoot immediately after each strain rate jump. This stress reaction to strain rate jumps is generally referred to as instantaneous strain rate sensitivity [5,13,14,38,60] and is usually followed by a transient period of stress before the flow stress reaches a new steady state. This instantaneous strain rate sensitivity always takes positive values and can be rationalized with the time-dependence of the solute concentration near mobile dislocations [36]. However, after reaching the steady state, the true stress-true strain curves exhibit a different behavior: In the range from 10^{-5} s^{-1} to 10^{-2} s^{-1} both materials actually respond with decreasing flow stresses when the strain rate is increased. Consequently, a decrease of the strain rate leads to an increase of the flow stress after each jump for both materials. From this behavior, a negative strain rate sensitivity for both materials in the W-condition is determined (see Table 2). For all jumps from low strain rates to high strain rates, negative strain rate sensitivity values m were calculated. With the negative strain rate sensitivity factors m for both materials a necessary requirement for the occurrence of PLC effects is fulfilled [27–30].

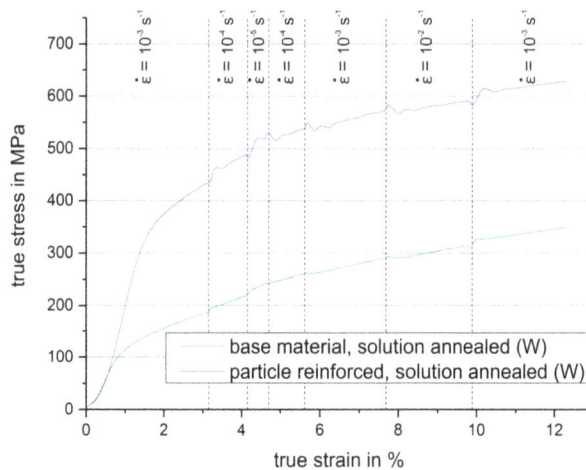

Figure 4. True stress-true strain curves measured in compressive jump tests with strain rates ranging from 10^{-5} s^{-1} to 10^{-2} s^{-1} for both materials in the solution annealed condition (W). When strain rates are reduced (first jumps and final jump), flow stresses are increased; an increase of the strain rate (intermediate jumps) leads to decreasing flow stresses. This clearly indicates that both materials exhibit a negative strain rate sensitivity.

Table 2. Strain rate sensitivity m, determined from different strain rate jumps for solution annealed AA 2017 (base material and particle reinforced alloy). All data represent are arithmetically averaged parameters ($\pm$standard deviations) from at least three similar tests.

Strain Rate Jump in s^{-1}	10^{-5} to 10^{-4}	10^{-4} to 10^{-3}	10^{-3} to 10^{-2}
True Strain in%	4.61	5.62	7.68
m base material	-0.0044 ± 0.0006	-0.0083 ± 0.0006	-0.0105 ± 0.0011
m reinforced	-0.0231 ± 0.0004	-0.0062 ± 0.0034	-0.0060 ± 0.0001

The results of additional jump tests with alternating strain rates (between $10^{-3}\,\text{s}^{-1}$ and $10^{-2}\,\text{s}^{-1}$) shown in Figure 5a confirms the negative strain rate sensitivity. Again, both materials exhibit the typical instantaneous (overshoot) strain rate sensitivity for all conditions (W and T4). The stress overshoot for the reinforced conditions is much higher than for the base material. Figure 5b shows the strain rate sensitivity values plotted versus true strain calculated from the jumps of stress-strain curves shown in Figure 5a. The results of both W-conditions again highlight that an increased strain rate leads to decreasing flow stresses. Interestingly, in case of both T4-conditions, a negative strain rate sensitivity was observed, too. However, the absolute *m*-values of these conditions are considerably smaller compared to the values determined for the W-conditions. We also note that absolute *m*-values exhibit a tendency to decrease with increasing plastic strain. This decrease is assumed to originate from aging during compressive testing as already reported in [2,4,5,32] (and for AMCs in [9]). Moreover, a decrease of absolute values of negative strain rate sensitivity is in good agreement with the absence of serrated flow (PLC effect) for the naturally aged condition (T4) in the tensile tests at RT mentioned above.

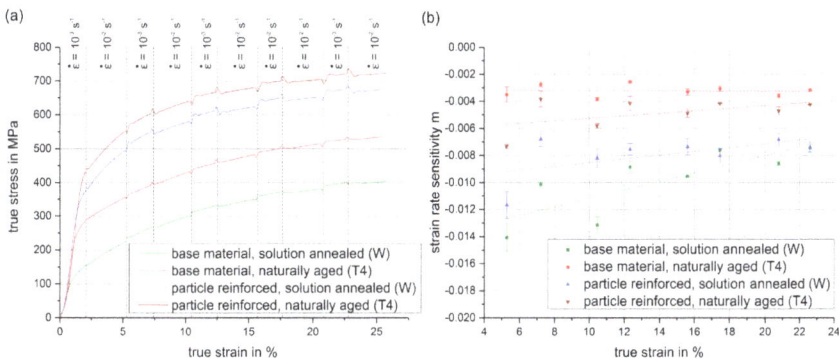

Figure 5. Compression tests with alternating strain rates between $10^{-3}\,\text{s}^{-1}$ and $10^{-2}\,\text{s}^{-1}$. (**a**) True stress-true strain curves of jump tests for both materials in solution annealed (W) and naturally aged condition (T4). (**b**) Strain rate sensitivity *m* versus true strain calculated from the true stress-true strain curves in a).

For the discussion of the influence of temperature on PLC effects, the results of the tensile tests performed at $-60\,^\circ\text{C}$ and $-196\,^\circ\text{C}$ and at a constant strain rate ($10^{-3}\,\text{s}^{-1}$) of the base material (in conditions W and T4) are presented in Figure 6. Figure 6a shows a reduction of stress serrations for the solution annealed condition W compared to the tensile tests at RT (see Figure 3a). Additionally, the critical strain for the onset of serrated flow in the W-condition is shifted to a larger engineering strain value of 11.3%. The T4-condition of the base material does not exhibit any serrations at $-60\,^\circ\text{C}$. At a testing temperature of $-196\,^\circ\text{C}$ (Figure 6b) the stress-strain curve of the W-condition does not show pronounced serrations; a slight waviness of the stress signal is most likely related to film boiling of the liquid nitrogen. A critical strain for the onset of stress serrations cannot be determined reliably for the W-condition. For the T4-condition, again, no serrations are observed.

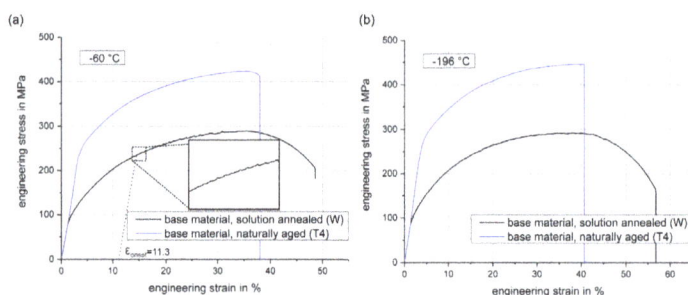

Figure 6. Tensile engineering stress-strain curves measured at low temperatures. (**a**) AA2017 base material in solution annealed condition W and in naturally aged condition T4 at −60 °C. (**b**) AA2017 base material in solution annealed condition W and in naturally aged condition T4 at −196 °C. Only the W-conditions still exhibit a slightly serrated flow associated with the PLC effect (inset at higher magnification). The serrations decrease with decreasing testing temperature.

Figure 7 shows the results of low-temperature tensile testing of the particle reinforced material for both conditions (W and T4). As already found for the base material (see Figure 6), the particle reinforced material in Figure 7a shows a significant decrease of serrated flow at −60 °C compared to RT (see Figure 3b). Again, the critical strain for the onset of stress serrations in the W-condition is shifted to larger strain values of about 7.0% compared to 1.9% at RT. At −60 °C, stress serrations are already completely suppressed for the T4-condition. Figure 7b shows a similar characteristic of stress-strain behavior compared to the behavior of the base material shown in Figure 6b. Lower testing temperatures (−196 °C) lead to a reduction of serrations in the stress-strain curve for the W-condition compared to testing at RT (see Figure 3b). Again, the critical strain for the onset of stress serrations cannot be determined clearly. For the T4-condition, an absence of serrations is observed, i.e., the PLC effect is completely suppressed. In summary, the suppression of serrated flow at lower temperatures (−60 °C and −196 °C, Figures 6 and 7) clearly indicates that thermally activated and diffusion-controlled processes are primarily responsible for serrated flow at RT in the materials studied here, irrespective of whether they contain particles (which tend to increase stress amplitudes during serrated flow) or not. Temperature reduction leads to reduced diffusivity of the solute atoms and thus to a lower efficiency of dynamic strain aging processes in the AA2017 alloy.

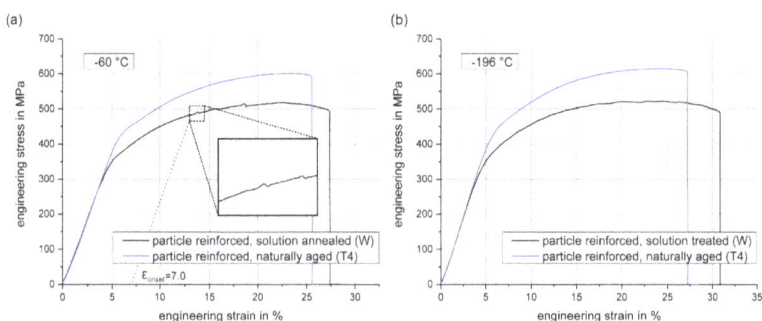

Figure 7. Tensile engineering stress-strain curves at low temperatures. (**a**) Particle reinforced AA2017 material in the solution annealed (W) and naturally aged (T4) conditions at −60 °C. (**b**) Particle reinforced AA2017 material in solution annealed (W) and naturally aged (T4) conditions at −196 °C. Only the W-conditions exhibit some serrated flow (inset at higher magnification). PLC effects are reduced by decreasing the testing temperature.

Additional experiments were performed to further study the material condition that exhibits the most pronounced PLC effect, particle reinforced AA2017 in W-condition. Simultaneous DIC and acoustic emission measurements were performed during uniaxial tensile tests at RT. Representative results of these tensile tests are shown in Figure 8. Figure 8a shows the evolution of stress, strain and of the acoustic emission signal as a function of time. In this figure, "macroscopic" strain represents the average uniaxial strain value across the entire surface area that was evaluated by DIC. Both stress and strain signals clearly indicate serrated flow; stress drops coincide with positive jumps of macroscopic strain. The acoustic emission signal is represented by the black line in Figure 8a (the signal simply represents the integral acoustic intensity). Each distinct stress drop is also accompanied by perceptible acoustic emission, see also the video provided as supplementary material (Video S1: PLC effect reinforced AA2017 using DIC and AE). The DIC observations show that the measured acoustic emission is clearly associated with the nucleation of a separate deformation band in the material, which is accompanied by a distinct stress drop at the same moment as band nucleation provides a sudden increase in local strain and thus partially elastically unloads the tensile sample. In Figure 8b, the nucleation and two stages during the propagation of one representative deformation band are shown as DIC-derived strain maps, where strain values correspond to uniaxial tensile strains determined in the (loading) *x*-direction. These characteristic situations are marked in the stress curve of Figure 8a (A,B,C). The nucleation (A) of a single deformation band at the lower end of the gauge length accompanied by a distinct stress drop can be clearly identified. Further deformation proceeds by propagation (B) of the single band across the gauge length. This is accompanied by very small serrations of the stress signal, most likely due to dislocation particle interactions at the reaction front of the deformation band [42]. Interestingly, no secondary PLC band nucleation is observed, confirming that the PLC effect observed here is not of type B. Immediately prior to the next distinct stress drop (accompanied by the nucleation of a new deformation band) the current PLC band reaches the opposite end of the gauge length (C). This process occurs repeatedly during tensile deformation until final fracture. The combination of DIC and acoustic emission provides detailed information on nucleation events and on the propagation of individual PLC bands. Acoustic emission may even contribute to an analysis of the microstructural interaction of the growing band with the material's reinforcement particles, which is the subject of ongoing work.

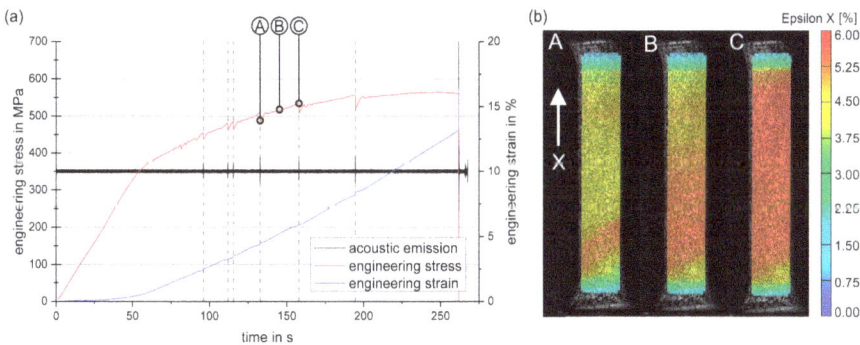

Figure 8. (**a**) Engineering stress, engineering strain and acoustic emission signal (the signal simply represents the integral acoustic intensity) as a function of testing time during uniaxial tensile testing of the particle reinforced material (W-condition) at RT. The stress signal (red) shows distinct stress-drops accompanied by strain discontinuities (blue curve) and a perceptible acoustic emission (black curve). (**b**) Strain fields measured by DIC (uniaxial tensile strain maps) at different stages of the evolution of a single PLC band (type A): Nucleation (A) and propagation (B to C) of a single deformation band.

4. Summary and Conclusions

Serrated flow was investigated in an age-hardenable Al-Cu alloy AA2017 and in a particle reinforced AA2017 alloy (10 vol. % SiC) as a function of heat treatment condition, applied strain rate, as well as testing temperature. Two different heat treatment conditions were analyzed: solid solution annealed (W) and naturally aged (T4). It was found that both materials exhibit a typical serrated flow in the W-condition during tensile testing at ambient temperature. Furthermore, a difference in the plastic strain onset of the initiation of serrated flow was observed. An earlier onset occurs in the particle reinforced material, which can be attributed to an accelerated dynamic strain aging due to a comparatively faster increase of dislocation density during plastic deformation. We also documented a negative strain rate sensitivity during compressive jump tests from 10^{-5} s^{-1} to 10^{-2} s^{-1} for both materials and both conditions. The (negative) strain rate sensitivity values determined for the T4-conditions show considerably lower absolute values compared to the W-condition. This is in line with the observation of a reduced serrated flow (PLC effect) for the naturally aged condition (T4) in tensile tests at RT.

During tensile tests at lower deformation temperatures (-60 °C, -196 °C), a shift of the critical strain values for the onset of serrated flow in the W-condition to larger strain values and a decreased amplitude of stress fluctuations were observed. Serrated flow is completely suppressed at lower temperatures for T4-condition. The temperature-dependence of serrated flow in both materials reveals that thermally activated and diffusion-controlled processes are the main reason for the different flow behavior of the materials at RT, -60 °C, and -196 °C. The nucleation and propagation of single PLC bands were characterized using DIC and acoustic emission measurements for the condition with the most pronounced PLC effect (particle reinforced AA2017 in W-condition). DIC strain maps confirm that the PLC bands are of type A. Acoustic emission data can be directly related to nucleation events of distinct bands.

Supplementary Materials: The following are available online at www.mdpi.com/2075-4701/8/2/88/s1, Video S1: PLC effect reinforced AA2017 using DIC and AE.

Acknowledgments: The authors gratefully acknowledge funding by the German Research Foundation (Deutsche Forschungsgemeinschaft, DFG) through the Collaborative Research Center SFB 692 (projects A2 and C5). We also thank our partners in project A2, Daisy Julia Nestler, and Steve Siebeck, for fabrication of the reinforced materials.

Author Contributions: Markus Härtel and Christian Illgen conceived and designed the experiments. Christian Illgen performed and analyzed the experiments, prepared the figures. Markus Härtel and Christian Illgen drafted the manuscript. Philipp Frint and Martin Franz-Xaver Wagner discussed the results and analysis and helped writing the manuscript.

Conflicts of Interest: The authors declare no conflict of interest.

References

1. Böhlke, T.; Bondár, G.; Estrin, Y.; Lebyodkin, M.A. Geometrically non-linear modeling of the Portevin-Le Chatelier effect. *Comput. Mater. Sci.* **2009**, *44*, 1076–1088. [CrossRef]
2. Chmelík, F.; Pink, E.; Król, J.; Balík, J.; Pešička, J.; Lukáč, P. Mechanisms of serrated flow in aluminium alloys with precipitates investigated by acoustic emission. *Acta Mater.* **1998**, *46*, 4435–4442. [CrossRef]
3. Delaunois, F.; Denil, E.; Marchal, Y.; Vitry, V. Accelerated aging and Portevin-Le Chatelier effect in AA 2024. *Mater. Sci. Forum* **2017**, *879*, 524–529. [CrossRef]
4. Gupta, S.; Beaudoin, A.J.; Chevy, J. Strain rate jump induced negative strain rate sensitivity (NSRS) in aluminum alloy 2024: Experiments and constitutive modeling. *Mater. Sci. Eng. A* **2017**, *683*, 143–152. [CrossRef]
5. Jiang, H.; Zhang, Q.; Chen, X.; Chen, Z.; Jiang, Z.; Wu, X.; Fan, J. Three types of Portevin-Le Chatelier effects: Experiment and modelling. *Acta Mater.* **2007**, *55*, 2219–2228. [CrossRef]
6. Jiang, H.; Zhang, Q.; Wu, X.; Fan, J. Spatiotemporal aspects of the Portevin-Le Chatelier effect in annealed and solution-treated aluminum alloys. *Scr. Mater.* **2006**, *54*, 2041–2045. [CrossRef]

7. Jiang, Z.; Zhang, Q.; Jiang, H.; Chen, Z.; Wu, X. Spatial characteristics of the Portevin-Le Chatelier deformation bands in Al-4 at%Cu polycrystals. *Mater. Sci. Eng. A* **2005**, *403*, 154–164. [CrossRef]
8. Rashkeev, S.N.; Glazov, M.V.; Barlat, F. Strain-rate sensitivity limit diagrams and plastic instabilities in a 6xxx series aluminum alloy. Part I: Analysis of temporal stress-strain serrations. *Comput. Mater. Sci.* **2002**, *24*, 295–309. [CrossRef]
9. Serajzadeh, S.; Ranjbar Motlagh, S.; Mirbagheri, S.M.H.; Akhgar, J.M. Deformation behavior of AA2017-SiCp in warm and hot deformation regions. *Mater. Des.* **2015**, *67*, 318–323. [CrossRef]
10. Wagner, S.; Härtel, M.; Frint, P.; Wagner, M.F.X. Influende of ECAP on the formability of a particle reinforced 2017 aluminum alloy. *J. Phys. Conf. Ser.* **2017**, *755*, 11001. [CrossRef]
11. Xiang, G.F.; Zhang, Q.C.; Liu, H.W.; Wu, X.P.; Ju, X.Y. Time-resolved deformation measurements of the Portevin-Le Chatelier bands. *Scr. Mater.* **2007**, *56*, 721–724. [CrossRef]
12. Zhang, Q.; Jiang, Z.; Jiang, H.; Chen, Z.; Wu, X. On the propagation and pulsation of Portevin-Le Chatelier deformation bands: An experimental study with digital speckle pattern metrology. *Int. J. Plast.* **2005**, *21*, 2150–2173. [CrossRef]
13. Zhang, S.; McCormick, P.G.; Estrin, Y. The morphology of Portevin-Le Châtelier bands: Finite element simulation for Al-Mg-Si. *Acta Mater.* **2001**, *49*, 1087–1094. [CrossRef]
14. Dierke, H.; Krawehl, F.; Graff, S.; Forest, S.; Šachl, J.; Neuhäuser, H. Portevin-LeChatelier effect in Al-Mg alloys: Influence of obstacles-experiments and modelling. *Comput. Mater. Sci.* **2007**, *39*, 106–112. [CrossRef]
15. Chmelík, F.; Klose, F.B.; Dierke, H.; Šachl, J.; Neuhäuser, H.; Lukáč, P. Investigating the Portevin-Le Châtelier effect in strain rate and stress rate controlled tests by the acoustic emission and laser extensometry techniques. *Mater. Sci. Eng. A* **2007**, *462*, 53–60. [CrossRef]
16. Liang, S.; Qing-Chuan, Z.; Peng-Tao, C. Influence of solute cloud and precipitates on spatiotemporal characteristics of Portevin-Le Chatelier effect in A2024 aluminum alloys. *Chin. Phys. B* **2009**, *18*, 3500–3507. [CrossRef]
17. Lebedkina, T.A.; Lebyodkin, M.A. Effect of deformation geometry on the intermittent plastic flow associated with the Portevin-Le Chatelier effect. *Acta Mater.* **2008**, *56*, 5567–5574. [CrossRef]
18. Grzegorczyk, B.; Ozgowicz, W.; Kalinowska-Ozgowicz, E.; Kowalski, A. Investigation of the Portevin-Le Chatelier effect by the acoustic emission. *J. Achiev. Mater. Manuf. Eng.* **2013**, *60*, 7–14.
19. Cottrell, A.H. LXXXVI. A note on the Portevin-Le Chatelier effect. *Lond. Edinb. Dublin Philos. Mag. J. Sci.* **1953**, *44*, 829–832. [CrossRef]
20. Robinson, J.M.; Shaw, M.P. Microstructural and mechanical influences on dynamic strain aging phenomena. *Int. Mater. Rev.* **1994**, *39*, 113–122. [CrossRef]
21. Yilmaz, A. The Portevin-Le Chatelier effect: A review of experimental findings. *Sci. Technol. Adv. Mater.* **2011**, *12*, 063001. [CrossRef] [PubMed]
22. Rodriguez, P.; Venkadesan, S. Serrated Plastic Flow Revisited. *Solid State Phenom.* **1995**, *42–43*, 257–266. [CrossRef]
23. Rizzi, E.; Hähner, P. On the Portevin-Le Chatelier effect: Theoretical modeling and numerical results. *Int. J. Plast.* **2004**, *20*, 121–165. [CrossRef]
24. Le Chatelier, A. Influence du temps et de la température sur les essais au choc. *Rev. Met. Paris* **1909**, *6*, 914–917. [CrossRef]
25. Portevin, A.; Le Chatelier, F. Sur un phénomène observé lors de l'essai de traction d'alliages en cours de transformation. *Comptes Rendus Acad. Sci. Paris* **1923**, *176*, 507–510.
26. Considère, M. Memoire sur l'emploi du fer et de l'acier dans les constructions. *Ann. Ponts Chaussees* **1885**, *9*, 574–775.
27. Hähner, P. On the physics of the Portevin-Le Châtelier effect part 1: The statistics of dynamic strain ageing. *Mater. Sci. Eng. A* **1996**, *207*, 208–215. [CrossRef]
28. Hähner, P. On the physics of the Portevin-Le Chgttelier effect part 2: From microscopic to macroscopic behaviour. *Mater. Sci. Eng. A* **1996**, *207*, 216–223. [CrossRef]
29. Hähner, P.; Ziegenbein, A.; Rizzi, E.; Neuhäuser, H. Spatiotemporal analysis of Portevin–Le Châtelier deformation bands: Theory, simulation, and experiment. *Phys. Rev. B* **2002**, *65*, 134109. [CrossRef]
30. Bross, S.; Hähner, P.; Steck, E.A. Mesoscopic simulations of dislocation motion in dynamic strain ageing alloys. *Comput. Mater. Sci.* **2003**, *26*, 46–55. [CrossRef]

31. Kubin, L.P.; Estrin, Y. Evolution of dislocation densities and the critical conditions for the Portevin-Le Châtelier effect. *Acta Metall. Mater.* **1990**, *38*, 697–708. [CrossRef]
32. Fritsch, S.; Scholze, M.; Wagner, M.F.X. Influence of thermally activated processes on the deformation behavior during low temperature ECAP. *IOP Conf. Ser. Mater. Sci. Eng.* **2016**, *118*, 012030. [CrossRef]
33. Härtel, M.; Wagner, S.; Frint, P.; Wagner, M.F.X. Effects of particle reinforcement and ECAP on the precipitation kinetics of an Al-Cu alloy. *IOP Conf. Ser. Mater. Sci. Eng.* **2014**, *63*, 012080. [CrossRef]
34. Härtel, M.; Frint, P.; Abstoss, K.G.; Wagner, M.F.-X. Effect of Creep and Aging on the Precipitation Kinetics of an Al-Cu Alloy after One Pass of ECAP. *Adv. Eng. Mater.* **2017**, *1700307*, 1700307. [CrossRef]
35. Wagner, S.; Siebeck, S.; Hockauf, M.; Nestler, D.; Podlesak, H.; Wielage, B.; Wagner, M.F.X. Effect of SiC-reinforcement and equal-channel angular pressing on microstructure and mechanical properties of AA2017. *Adv. Eng. Mater.* **2012**, *14*, 388–393. [CrossRef]
36. McCormick, P.G. Theory of flow localisation due to dynamic strain ageing. *Acta Metall.* **1988**, *36*, 3061–3067. [CrossRef]
37. Balík, J.; Lukáč, P.; Kubin, L.P. Inverse critical strains for jerky flow in Al-Mg alloys. *Scr. Mater.* **2000**, *42*, 465–471. [CrossRef]
38. Picu, R.C.; Vincze, G.; Ozturk, F.; Gracio, J.J.; Barlat, F.; Maniatty, A.M. Strain rate sensitivity of the commercial aluminum alloy AA5182-O. *Mater. Sci. Eng. A* **2005**, *390*, 334–343. [CrossRef]
39. Schmahl, W.W.; Khalil-Allafi, J.; Hasse, B.; Wagner, M.; Heckmann, A.; Somsen, C. Investigation of the phase evolution in a super-elastic NiTi shape memory alloy (50.7 at.%Ni) under extensional load with synchrotron radiation. *Mater. Sci. Eng. A* **2004**, *378*, 81–85. [CrossRef]
40. Cai, Y.; Zhang, Q.; Yang, S.; Fu, S.; Wang, Y. Experimental Study on Three-Dimensional Deformation Field of Portevin-Le Chatelier Effect Using Digital Image Correlation. *Exp. Mech.* **2016**, *56*, 1243–1255. [CrossRef]
41. Schäfer, A.; Wagner, M.F.-X.; Pelegrina, J.L.; Olbricht, J.; Eggeler, G. Localization events and microstructural evolution in ultra fine grained NiTi shape memory alloys during thermo-mechanical loading. *Adv. Eng. Mater.* **2010**, *12*, 453–459. [CrossRef]
42. Delpueyo, D.; Balandraud, X.; Grédiac, M. Calorimetric signature of the Portevin-Le Châtelier effect in an aluminum alloy from infrared thermography measurements and heat source reconstruction. *Mater. Sci. Eng. A* **2016**, *651*, 135–145. [CrossRef]
43. Elibol, C.; Wagner, M.F.-X. Investigation of the stress-induced martensitic transformation in pseudoelastic NiTi under uniaxial tension, compression and compression-shear. *Mater. Sci. Eng. A* **2004**, *621*, 76–81. [CrossRef]
44. Le Cam, J.-B.; Robin, E.; Leotoing, L.; Guines, D. Calorimetric analysis of Portevin-Le Chatelier bands under equibiaxial loading conditions in Al–Mg alloys: Kinematics and mechanical dissipation. *Mech. Mater.* **2017**, *105*, 80–88. [CrossRef]
45. Zdunek, J.; Płowiec, J.; Spychalski, W.L.; Mizera, J. Acoustic emission studies of the Portevin-Le Chatelier effect in Al-Mg-Mn (5182) alloy. *Inż. Mater.* **2011**, *32*, 889–894.
46. Chemelik, F.; Ziegenbein, A.; Neuhauser, H.; Lukac, P. Investigating the Portevin–Le Châtelier effect by the acoustic emission and laser extensometry techniques. *Mater. Sci. Eng. A* **2002**, *324*, 200–207. [CrossRef]
47. Wevers, M. Listening to the sound of materials: Acoustic emission for the analysis of material behaviour. *NDT E Int.* **1997**, *30*, 99–106. [CrossRef]
48. Estrin, Y.; Lebyodkin, M.A. The influence of dispersion particles on the Portevin-Le Chatelier effect: From average particle characteristics to particle arrangement. *Mater. Sci. Eng. A* **2004**, *387–389*, 195–198. [CrossRef]
49. Thevenet, D.; Mliha-Touati, M.; Zeghloul, A. The effect of precipitation on the Portevin-Le Chatelier effect in an Al-Zn-Mg-Cu alloy. *Mater. Sci. Eng. A* **1999**, *266*, 175–182. [CrossRef]
50. McCormick, P.G. A model for the Portevin-Le Chatelier effect in substitutional alloys. *Acta Metall.* **1972**, *20*, 351–354. [CrossRef]
51. Beukel, A. Theory of the effect of dynamic strain aging on mechanical properties. *Phys. Status Solidi* **1975**, *30*, 197–206. [CrossRef]
52. Mulford, R.A.; Kocks, U.F. New observations on the mechanisms of dynamic strain aging and of jerky flow. *Acta Metall.* **1979**, *27*, 1125–1134. [CrossRef]
53. Lebyodkin, M.; Dunin-Barkowskii, L.; Bréchet, Y.; Estrin, Y.; Kubin, L.P. Spatio-temporal dynamics of the Portevin-Le Chatelier effect: Experiment and modelling. *Acta Mater.* **2000**, *48*, 2529–2541. [CrossRef]

54. Staab, T.E.M.; Haaks, M.; Modrow, H. Early precipitation stages of aluminum alloys-The role of quenched-in vacancies. *Appl. Surf. Sci.* **2008**, *255*, 132–135. [CrossRef]
55. Sauvage, X.; Wilde, G.; Divinski, S.V.; Horita, Z.; Valiev, R.Z. Grain boundaries in ultrafine grained materials processed by severe plastic deformation and related phenomena. *Mater. Sci. Eng. A* **2012**, *540*, 1–12. [CrossRef]
56. Kim, W.J.; Chung, C.S.; Ma, D.S.; Hong, S.I.; Kim, H.K. Optimization of strength and ductility of 2024 Al by equal channel angular pressing (ECAP) and post-ECAP aging. *Scr. Mater.* **2003**, *49*, 333–338. [CrossRef]
57. Divinski, S.V.; Reglitz, G.; Rösner, H.; Estrin, Y.; Wilde, G. Ultra-fast diffusion channels in pure Ni severely deformed by equal-channel angular pressing. *Acta Mater.* **2011**, *59*, 1974–1985. [CrossRef]
58. Schwink, C.; Nortmann, A. The present experimental knowledge of dynamic strain ageing in binary f.c.c. solid solutions. *Mater. Sci. Eng. A* **1997**, *234–236*, 1–7. [CrossRef]
59. Chihab, K.; Estrin, Y.; Kubin, L.P.; Vergnol, J. The kinetics of the Portevin-Le Chatelier bands in an Al-5at%Mg alloy. *Scr. Metall.* **1987**, *21*, 203–208. [CrossRef]
60. Ling, C.P.; McCormick, P.G. Strain rate sensitivity and transient behaviour in an Al-Mg-Si alloy. *Acta Metall. Mater.* **1990**, *38*, 2631–2635. [CrossRef]

metals

MDPI

Article

Influence of Boron on the Creep Behavior and the Microstructure of Particle Reinforced Aluminum Matrix Composites

Steve Siebeck [1],*, Kristina Roder [1], Guntram Wagner [1] and Daisy Nestler [2]

[1]   Institute of Materials Science and Engineering, Chemnitz University of Technology, Chair of Composites and Material Compounds, D-09107 Chemnitz, Germany; kristina.roder@mb.tu-chemnitz.de (K.R.); guntram.wagner@mb.tu-chemnitz.de (G.W.)

[2]   Fakultät für Maschinenbau, Technische Universität Chemnitz, Professur Strukturleichtbau und Kunststoffverarbeitung, D-09107 Chemnitz, Germany; daisy.nestler@mb.tu-chemnitz.de

*   Correspondence: steve.siebeck@mb.tu-chemnitz.de; Tel.: +49-371-531-37531

Received: 15 December 2017; Accepted: 30 January 2018; Published: 6 February 2018

Abstract: The reinforcement of aluminum alloys with particles leads to the enhancement of their mechanical properties at room temperature. However, the creep behavior at elevated temperatures is often negatively influenced. This raises the question of how it is possible to influence the creep behavior of this type of material. Within this paper, selected creep and tensile tests demonstrate the beneficial effects of boron on the properties of precipitation-hardenable aluminum matrix composites (AMCs). The focus is on the underlying microstructure behind this effect. For this purpose, boron was added to AMCs by means of mechanical alloying. Comparatively higher boron contents than in steel are investigated in order to be able to record their influence on the microstructure including the formation of potential new phases as well as possible. While the newly formed phase Al_3BC can be reliably detected by X-ray diffraction (XRD), it is difficult to obtain information about the phase distribution by means of scanning electron microscopy (SEM) and scanning transmission electron microscopy (STEM) investigations. An important contribution to this is finally provided by the investigation using Raman microscopy. Thus, the homogeneous distribution of finely scaled Al_3BC particles is detectable, which allows conclusions about the microstructure/property relationship.

Keywords: aluminum matrix composites; creep behavior; Raman microscopy; Al_3BC; particle reinforced; powder metallurgy

1. Introduction

Within the Collaborative Research Center 692, a series of particle-reinforced aluminum matrix composites (AMCs) based on a precipitation-hardenable aluminum alloy were produced and investigated. The results on room temperature properties are published in [1–6]. In addition, creep tests were carried out, in which all tested AMCs showed higher creep rates than the unreinforced reference material.

An important mechanism that influences the creep rate of particle-reinforced materials is the acceleration of the precipitation kinetics [7]. Particularly, this is caused by the high dislocation density in the vicinity of the reinforcement particles [8]. The reference material was consolidated analogously to the AMCs by hot isostatic pressing and extrusion. However, it was not high-energy ball milled. This means that in addition to the particle reinforcement itself, the strong strain hardening is a significant difference between the AMCs and the reference material. This is in turn accompanied by the formation of a very fine-grained microstructure. Unlike at the time-independent plastic deformation, a fine-grained structure has a negative effect on the creep behavior. For example, in case of diffusional creep, the creep rate behaves in inverse proportion to the square of the grain size and in case of grain boundary diffusion even to the cube [9,10]. Furthermore, it was confirmed that the direct effects of the particles compared to the indirect ones by influencing the matrix material is smaller [11].

This correlation was also confirmed by a series of experiments on the creep behavior of the highly plastically deformed reference material by equal channel angular pressing (ECAP) [12]. From this, it can be derived that in addition to grain boundary sliding, diffusion-controlled mechanisms in particular dominate the creep behavior of the investigated AMCs. These include the mechanisms of creep as well as the accelerated precipitation kinetics. Thus, a possible approach to increase the resistance against creep shall be the obstruction of diffusion processes or the stabilization of the grain boundaries. In steels, this is realized by alloying elements with a small atomic radius, such as boron or lithium [13,14].

The resistance against creep is already significantly increased by a boron content in the ppm range [15]. Boron dissolved in austenite has a high binding energy to lattice defects such as dislocations and vacancies and therefore segregates at grain boundaries during cooling [16,17]. This reduces both the grain boundary energy and the diffusivity for iron and carbon along the grain boundaries. Depending on the steel, the precipitation of $M_{23}(B,C)_6$ phases (borocarbides) with epitaxies to the adjacent austenite grains follows. They stabilize the grain boundaries and are able to close cavities [18,19]. The formation of borocarbides may also be considered an indicator of excessive boron contents [20]. No indication can be found in the literature for a specific influence on the creep behavior of aluminum materials by boron. The behavior of the material system aluminum and boron by means of mechanical alloying has already been investigated [21]. When using up to 50 wt % B, the formation of the boride AlB_{10} could be observed.

Within this paper, selected creep and tensile tests demonstrate the beneficial effects of boron on the properties of precipitation-hardenable AMCs. The focus is on the underlying microstructure behind this effect. For this purpose, boron was added to micro-scaled as well as to a nano-scaled SiCp-reinforced AMC. Higher boron contents than in steel are investigated in order to be able to record the influence on the microstructure including the formation of potential new phases. While the newly formed phase Al_3BC can be reliably detected by X-ray diffraction (XRD), it is difficult to obtain optical information about the phase distribution by means of scanning electron microscopy (SEM) and scanning transmission electron microscopy (STEM). An important contribution to this is finally provided by the investigation using Raman microscopy. Thus, the homogeneous distribution of finely scaled Al_3BC particles can be detected, which allows conclusions about the microstructure/property relationship.

2. Materials and Methods

The aluminum alloy that was used as matrix material was supplied in the form of a commercial, gas-atomized, spherical powder with a particle size fraction <100 µm (TLS Technik GmbH & Co. Spezialpulver KG, Bitterfeld, Germany). The chemical composition (in weight-percent) of the alloy was 3.9% Cu, 0.7% Mg, 0.6% Mn, 0.1% Si, 0.2% Fe, balance Al (AA2017). As small particle sizes minimize local stress concentrations in the aluminum matrix, fine-grained SiC alpha phase powder with a fraction of smaller than 1 µm (97.5%, ESK-SIC GmbH, Frechen, Germany) as well as a nano sized beta phase with a fraction smaller than 200 nm (PlasmaChem, Berlin, Germany) were used as reinforcing components.

The composite powder was processed in a water-cooled high-energy ball mill Simoloyer® CM08 (Zoz Company, Wenden, Germany) with ceramic lining. Milling was performed for at least four hours in air atmosphere (closed milling chamber). The remaining oxygen is incorporated into the composite powder and this leads to the formation of finely dispersed spinels ($MgAl_2O_4$) in the bulk material. As the spinels are finely distributed in rather smaller quantities, there is no negative effect expected [22]. Since the powder has non-passivated surfaces after the milling process, there is a possibility of heating. This happens only with an intense supply of oxygen and the reaction is generally moderate due to the size of the composite particles. However, appropriate precautions should be taken into consideration. Results regarding the influence of the milling atmosphere were published in [2].

The addition of boron was also done by means of high-energy ball milling. Boron particles (>95%, Sigma-Aldrich, St. Louis, MO, USA) with a size of approx. 1 μm have been used for this purpose. The boron particles behave similarly to the ceramic particles within the aluminum matrix. To limit welding of the particles, as well as adhesion to the rotor, balls, and chamber, small amounts of stearic acid (C18H36O2, Merck KGaA, Darmstadt, Germany) were added as process control agent [23]. To remove the stearic acid after milling, the composite powder was first subjected to hot-degassing (450 °C, 6 × 10^{-2} bar, 4 h). Compaction for all materials was then performed by hot isostatic pressing at 450 °C for 3 h and at a pressure of 1100 bar. Finally, the material was extruded in a temperature range between 355 and 370 °C to produce semi-finished square bars with a cross section of 15 × 15 mm^2. The extrusion was performed with a punch speed of 2 mm/s and an extrusion ratio of 42:1. Hot isostatic pressing and extrusion were realized by PLM GmbH (Gladbeck, Germany). All materials were solid solution heat treated at 505 °C for at least 60 min, water quenched and naturally aged at room temperature (T4 condition). Details on the preparation of this kind of AMCs are published in [3–5,22,24]. Table 1 shows the composition of the five AMCs that are the subject of this paper.

Table 1. Composition of the investigated AMCs (aluminum matrix composites).

Volume Fraction in vol %		
SiC < 1 μm	SiC < 200 nm	Boron
10	-	-
10	-	0.9
10	-	3.0
-	5	-
-	5	5.0

For the characterization of strength and ductility, cylindrical tensile specimens (with an aspect ratio of the gauge length of three) were machined from the billets in the direction of extrusion. Quasi-static tensile tests were performed at room temperature in a conventional testing machine (Zwick GmbH & Co. KG, Ulm, Germany) with a constant cross-head speed corresponding to an initial strain rate of 10^{-3} s^{-1}. At least three tests were performed for the different material conditions, in particular to provide better statistics on fracture strains of the AMCs.

The creep tests were carried out on a creep testing machine ATS 2330 (ATS Inc., Butler, PA, USA) with a maximum force of 53 kN. All tests were carried out under uniaxial tensile load. The sample geometry was cylindrical with a measuring length of 30 mm and a measuring length diameter of 6 mm. The total length of the samples was 80 mm. They were fastened within the testing machine by means of two ISO metric screw threads (M12). The qualitative creep tests were carried out at a temperature of 180 °C. This corresponds to a homologous temperature of 0.35 of the matrix alloy, which is the maximum service temperature of the AA2017 alloy. The tension used in all experiments was 200 MPa. At least two parallel samples were tested in each case.

Metallographic sections of the bulk material were prepared by mechanical grinding and polishing. The microstructures were analyzed by scanning electron microscopy (SEM) and scanning transmission electron microscopy (STEM) on thin sections using a Neon40 (Zeiss AG, Oberkochen, Germany) field emission microscope operating at 30 kV with energy dispersive X-ray spectroscopy (EDX). The specimens for optical micrograph (OM) analysis were taken from the transverse plane and from the flow plane (i.e., parallel to the direction of extrusion). To determine the phase composition, XRD investigations were performed with a D8 Discover (Bruker Corp., Billerica, MA, USA). The measurements were carried out using a copper K-alpha radiation.

For the additional high-resolution chemical analysis, selected samples were analyzed by flight of time secondary ions mass spectrometry (ToF-SIMS) by tascon GmbH (Münster, Germany).

The minimum diameter of the beam was 300 nm. By recording a spectrum for each sample point, information about the atomic and molecular structure of the outer 1 to 3 monolayers of the solid was obtained. The sensitivities are in the ppm range with a lateral resolution of up to 100 nm.

Raman studies were performed using the inVia Raman microscope (Renishaw, Gloucestershire, UK) with a frequency-doubled Nd:YAG laser having a wavelength of 532 nm, and a maximum power of 100 mW. The advantage of Raman microscopy or Spectroscopy over the X-ray fine structure analysis is that using Raman many spectra within an area of a few 100 μm^2 can be recorded due to scanning with a high spatial resolution. For this purpose, a laser power of 50%, a measuring time per step of 1 s, and a step size of 0.2 μm has been used. The spectra can then be assigned to specific phases based on the characteristic peaks. Finally, using this data, a spatially resolved mapping of the identified phases can be generated. Above all, not only ceramics can be measured using Raman but also intermetallic phases, in contrast to pure metals can be Raman-active. Accordingly, this method provides a useful complement to the microstructural characterization of AMC materials whenever the expected phases and associated Raman spectra are known. The SiC polytype can also be identified by Raman spectroscopy because of its characteristic peaks [25]. This includes the cubic and hexagonal modifications used within this paper.

3. Results and Discussion

3.1. Influene of Boron on the Creep Behavior and Tensile Strength

A boron content of 0.9 vol % slows down the minimum creep rate of the micro-scaled AMCs with 10 vol % volume fraction and prolongs the time to fracture, Figure 1a. Higher boron contents of 3.0 vol % and 5.0 vol % lead to a further marked improvement. It should be noted that the material with a boron content of 5.0 vol % is based on nanoscaled reinforced AMCs (5 vol % SiC particles). However, the reference AMCs without boron achieve similar results in the creep test.

The addition of boron also led to significant improvements in the stress-strain behavior, Figure 1b. Thus, a boron content of 0.9 vol % lead to a similar effect as caused by the increase of the SiC content by about 5 vol % in AMCs. The elongation at break decreased analogously. The further increase to 3.0 vol % boron caused only a relatively small increase in strength but in proportion to lower elongation.

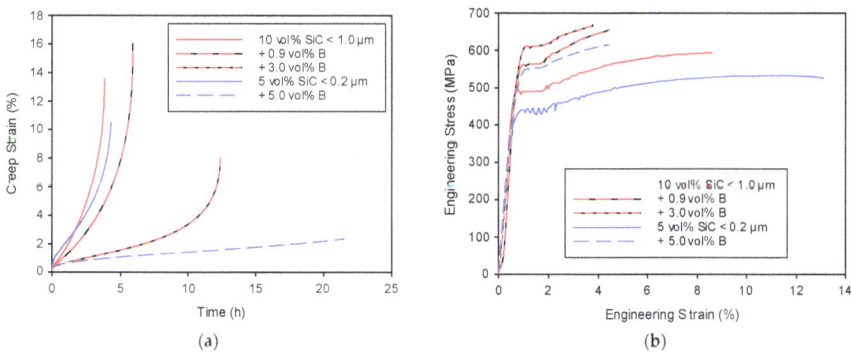

Figure 1. Influence of boron content on selected mechanical properties, based on: (**a**) tensile creep tests at a temperature of 180 °C and a stress of 200 MPa; (**b**) tensile tests.

3.2. Microstructure

In the extruded AMC material (Figure 2) boron-containing phases in the micron range can be detected. Especially on the ion-etched cross sections even smaller ones become visible. In Figure 2b, a boron enrichment at high-angle grain boundaries is possibly recognizable. By means of STEM, the identification of the boron containing particles succeeds only to a limited extent. There are now even more different phase particles with very small dimensions than in the AMC without boron. Figure 3 confirms the presence of boron by means of STEM on a rather large particle. The figure further illustrates the problem of the phase assignment in the present microstructure.

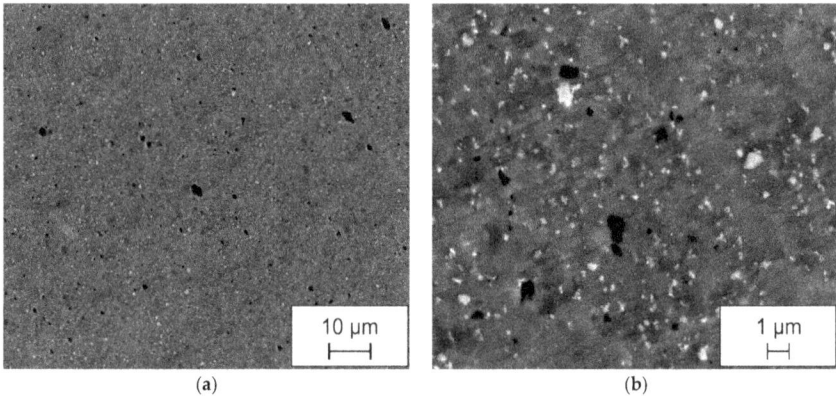

(a)   (b)

Figure 2. SEM images (SE) on the ion-etched cross section of an AMC reinforced with 10 vol % SiC and 3 vol % boron; matrix material: AA2017; size of the SiC and B particles: about 1 µm; B-containing phases (black), IM phases (white), SiC particles (hardly contrasted); (**a**,**b**) with different magnification.

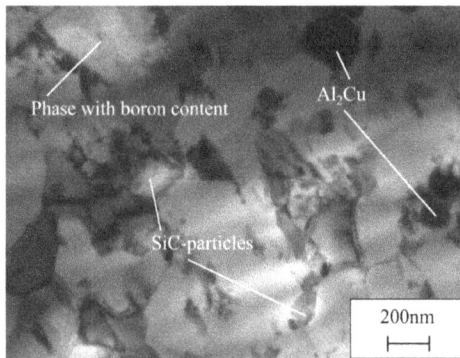

Figure 3. STEM (scanning transmission electron microscopy) image on the cross section of an AMC reinforced with 10 vol % SiC and 3 vol % boron.

In addition to the different structural components, TEM-typical diffraction contrast makes the evaluation of TEM and STEM examinations more difficult. These include different orientation, oblique grain boundaries and interfaces, crystal bendings, and the variation of the preparation thickness. Accordingly, it is probable that in addition to the proven phase, there may possibly be other considerably smaller boron-containing particles. ToF-SIMS investigations seem to be confirming the existence of a fine boron network at the grain boundaries (Figure 4). However, the clear evidence is still missing. The images also suggest that there is an affinity between B and Mg, which leads to the formation of Mg-B particles, presumably MgB_2. By comparison with the results of ToF-SIMS investigations on high energy milled powders, it can be demonstrated that this intermetallic Mg- and B-containing phase is formed at a higher temperature during the powder metallurgy processing and is not yet detectable in the powder stage (Figure 5).

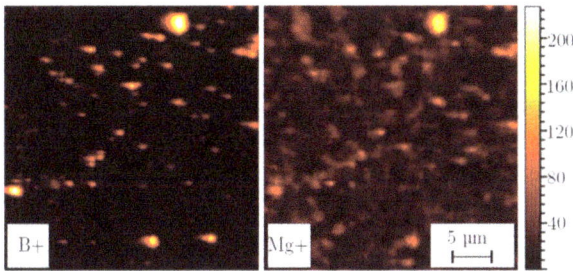

Figure 4. ToF-SIMS (flight of time secondary ions mass spectrometry) mapping on the cross section of an AMC reinforced with 10 vol % SiC and 3 vol % boron; matrix material: AA2017; size of the SiC and B particles: about 1 μm.

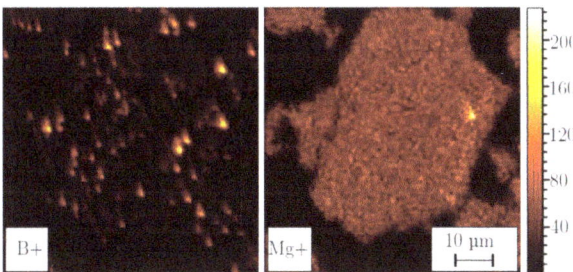

Figure 5. ToF-SIMS mapping on the powder section of composite powder reinforced with 10 vol % SiC and 3 vol % boron; matrix material: AA2017; size of the SiC and B particles: about 1 μm.

A reliable determination of the chemical composition by means of EDX is not possible due to the small size of the phase particles. Due to the low quantity and the low content, detection by XRD examinations is also not successful. However, the aluminum boron carbide phase Al$_3$BC, which was not yet been found in the powder is detectable (Figure 6).

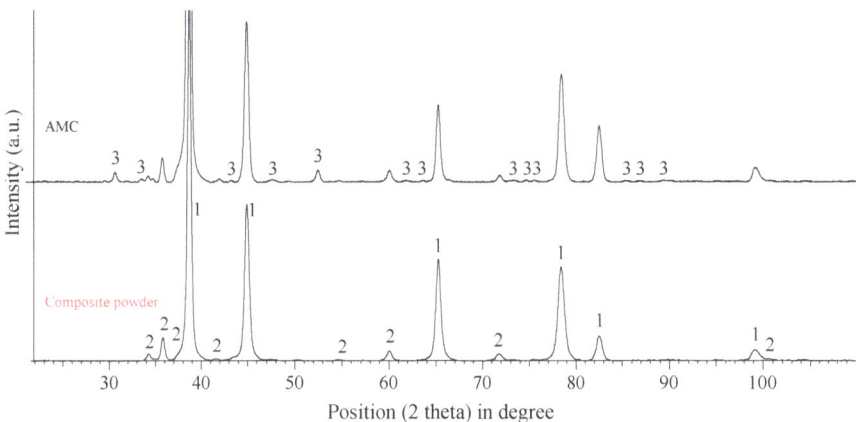

Figure 6. X-ray fine structure analysis of an AMC reinforced with 10 vol % SiC and 3 vol % boron and the underlying composite powder; matrix material: AA2017; size of the SiC and B particles: about 1 μm; 1: Al, 2: SiC, 3: Al$_3$BC.

The existence of this phase within the samples was confirmed by means of Raman microscopy. The basis for this investigation is the work of Meyer [26,27], who dealt with Raman and infrared spectra of Al_3BC and other ternary aluminum carbides. Figure 7 shows the results of the Raman investigations. The spectra of a sample containing 3 vol % of boron and the corresponding reference AMC without boron are obtained by averaging all the measured spectra within the mapping area. The band used for the evaluation of the false color representation is marked with the corresponding color. The identification of the phase Al_3BC succeeds from the bands detected by Meyer [26,28] at 147, 335, and 520 cm^{-1}.

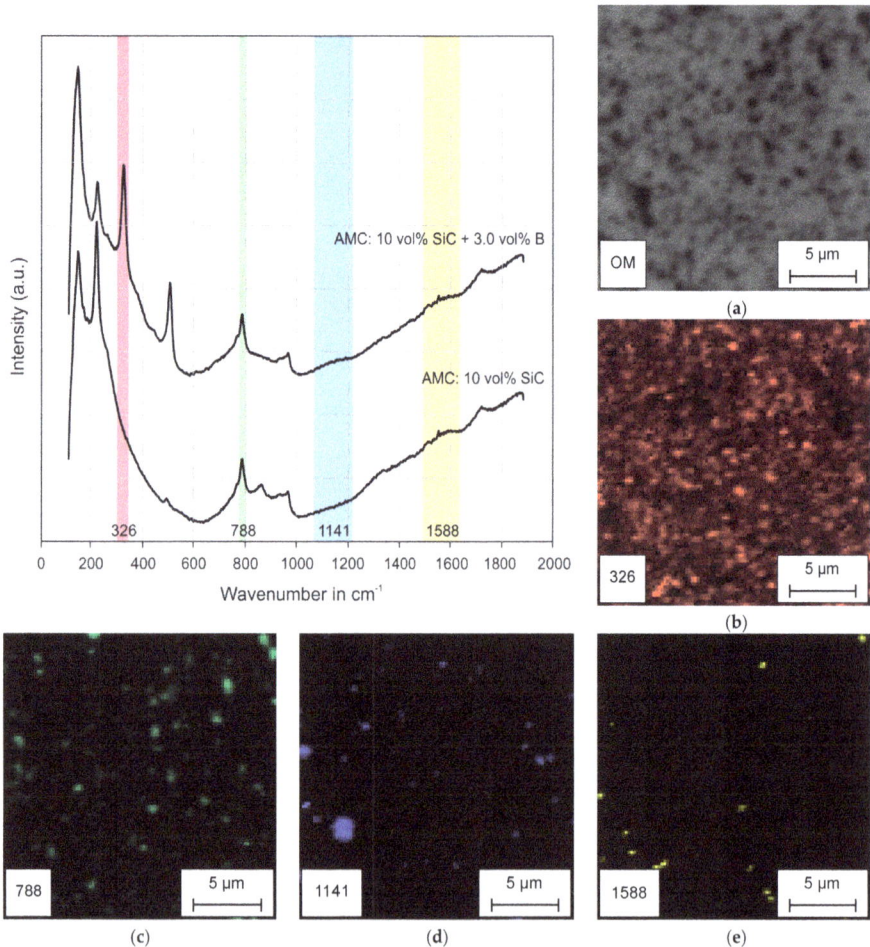

Figure 7. Raman mapping at an AMC with 10 vol % SiC and 3 vol % B and the averaged single spectra of the sample and the reference AMC with 10 vol % SiC; (a) optical micrograph of measuring area for mapping by bands: (b) at 326 cm^{-1} for Al_3BC, (c) at 788 cm^{-1} for SiC, (d) at 1141 cm^{-1} for boron, and (e) 1588 cm^{-1} for carbon.

The measured bands of the own samples are at 147, 326, and 508 cm^{-1}. This means that with increasing wave numbers they are slightly shifted towards smaller wave numbers. Such peak shifts indicate lattice distortions. In [29], the pressure-dependence of such shifts was investigated and illustrated by means of the Al_3BC_3 phase. Against this background, the peak shifts appears to be quite plausible with regard to the processing history of the AMCs. Further, in contrast to the finely divided phase particles, which apparently exist in a nanoscale, Meyer [26] recorded the spectra of single crystals with a size of $4 \times 4 \times 0.3$ mm^3. The Al_3BC mapping (red) illustrates the fine distribution by the fact that the selected band is detectable at almost every measuring point of the sample. In addition to Al_3BC, the bands of SiC, boron (B band for comparison in [30]) and carbon were evaluated. The latter is only very isolated and local, but already detectable in unreinforced samples, which was only hot isostatic pressed. Their detectability illustrates the already indicated presence of small amounts of carbon within the milled materials.

Nevertheless, even a small chemical interaction with the SiC particles cannot be ruled out. If the stabilization of the grain boundaries by boron atoms is to be the decisive mechanism of action, the explanation of this behavior would be that the higher boron content leads to a better boron distribution within the material. The fact that steels with significantly lower boron concentrations already produce the desired effect may be due to the atomic distribution at the grain boundaries. For the investigated AMCs, the only possible way to get a good boron distribution at the grain boundaries is through the diffusion processes that occur during powder metallurgical processing and heat treatment. A higher boron content would thus ultimately improve the distribution along the grain boundaries due to shorter diffusion paths and larger concentration differences. On the other hand, the excess of boron content obviously leads to the formation of sufficiently small, well-distributed borocarbide phases. This is the reason why neither the stress-strain behavior nor the creep behavior at elevated temperature is adversely affected. Like any particle reinforcement, they can also have a significant contribution to the creep behavior. This is particularly true if they, unlike precipitations, are thermally stable and thus remain very small.

4. Conclusions

Mechanical alloying has been used to add boron to particle-reinforced aluminum matrix composites. The main goal was to influence the creep behavior towards smaller creep rates. The results of the subsequent investigation of the creep behavior, the tensile strength, and the microstructure lead to the following conclusions:

- The distribution of boron particles succeeds. However, occasionally larger B-particles remain detectable. SEM and ToF-SIMS investigations indicate a fine distribution within the matrix.
- Both the creep resistance as well as the tensile strength increase with increasing boron content within the frame tested.
- By X-ray diffractometry, the newly formed phase Al_3BC can be detected. The existence of the phase can be confirmed by investigations using Raman microscopy. In addition, this makes it possible to visualize the distribution of the phase within the matrix by means of Raman mapping.
- The mechanism behind the improvement in creep resistance is not clear according to the current knowledge. The element boron may affect the diffusivity of the AMCs as well as grain boundary sliding as expected. On the other hand, it is also likely that the phase Al_3BC has a share in the improvement of the tested mechanical properties. After all, it is very small, well distributed, and stable against aging.

Acknowledgments: The authors would like to thank the Deutsche Forschungsgemeinschaft (DFG) for supporting the research project SFB 692 A2-1.

Author Contributions: Steve Siebeck is responsible for the production and characterization of the aluminum matrix composites. Kristina Roder accompanied the raman measurements with her professional expertise. Daisy Nestler and Guntram Wagner accompanied the material-technical part of the project and the manuscript with their professional expertise. All authors were involved in the preparation of the final version of the manuscript.

Conflicts of Interest: The authors declare no conflict of interest.

References

1. Hockauf, M.; Wagner, M.F.-X.; Händel, M.; Lampke, T.; Siebeck, S.; Wielage, B. High-strength aluminum-based light-weight materials for safety components-recent progress by microstructural refinement and particle reinforcement. *Int. J. Mater. Res.* **2012**, *103*, 3–11. [CrossRef]
2. Siebeck, S.; Nestler, D.; Podlesak, H.; Wielage, B. Influence of milling atmosphere on the high-energy ball-milling process of producing particle-reinforced aluminum matrix composites. In *Integration of Practice-Oriented Knowledge Technology: Trends and Prospectives*; Fathi, M., Ed.; Springer: Berlin/Heidelberg, Germany, 2013; pp. 315–321.
3. Siebeck, S.; Nestler, D.; Wielage, B. Producing a particle-reinforced AlCuMgMn alloy by means of mechanical alloying. *Materialwiss. Werkstofftech.* **2012**, *43*, 567–571. [CrossRef]
4. Siebeck, S.; Nestler, D.; Wielage, B. Hochenergie-Kugelmahlen zur Herstellung partikelverstärkter AMCs mit hochfester, maßgeschneiderter Aluminummatrix. In *Verbundwerkstoffe, Tagungsband*; Wanner, A., Ed.; Conventus Congressmanagement & Marketing: Jena, Germany, 2013; pp. 115–121.
5. Wagner, S.; Siebeck, S.; Hockauf, M.; Nestler, D.; Podlesak, H.; Wielage, B.; Wagner, M.F.-X. Effect of SiC-reinforcement and equal-channel angular pressing on microstructure and mechanical properties of AA2017. *Adv. Eng. Mater.* **2012**, *14*, 388–393. [CrossRef]
6. Nestler, D.J. *Beitrag zum Thema Verbundwerkstoffe-Werkstoffverbunde. Status quo und Forschungsansätze*; Universitätsverlag Chemnitz: Chemnitz, Germany, 2014.
7. Broeckmann, C.; Packeisen, A. Kriechen einer partikelverstärkten Aluminumlegierung unter Berücksich-tigung des Gefüges. *Metall* **1998**, *52*, 702–711.
8. Dutta, B.; Surappa, M.K. Age-hardening behaviour of Al-Cu-SiCp composites synthesized by casting route. *Scr. Metall. Mater.* **1995**, *32*, 731–736. [CrossRef]
9. Rösler, J.; Harders, H.; Bäker, M. *Mechanisches Verhalten der Werkstoffe*, 4; überarb. und erw. Aufl; Springer: Wiesbaden, Germany, 2012.
10. Bürgel, R. *Handbuch Hochtemperatur-Werkstofftechnik. Grundlagen, Werkstoffbeanspruchungen, Hochtempera-Turlegierungen und-Beschichtungen; mit 70 Tabellen*, 3; überarb. und erw. Aufl; Vieweg: Wiesbaden, Germany, 2006.
11. Rösler, J.; Bao, G.; Evans, A.G. The effects of diffusional relaxation on the creep strength of composites. *Acta Metall. Mater.* **1991**, *39*, 2733–2738. [CrossRef]
12. Härtel, M.; Frint, P.; Abstoss, K.G.; Wagner, M.F.-X. Effect of creep and aging on the precipitation kinetics of an Al-Cu Alloy after one pass of ECAP. *Adv. Eng. Mater.* **2017**, *20*. [CrossRef]
13. Melloy, G.F.; Slimmon, P.R.; Podgursky, P.P. Optimizing the boron effect. *Metall. Trans.* **1973**, *4*, 2279–2289. [CrossRef]
14. Choi, Y.-S.; Kim, S.-J.; Park, I.-M.; Kwon, K.-W.; Yoo, I.-S. Boron distribution in a low-alloy steel. *Met. Mater.* **1997**, *3*, 118–124. [CrossRef]
15. Kim, B.; Yun, H.; Lee, D.; Lim, B. Effect of boron on creep characteristics in 9Cr-1.5Mo alloys. *J. Phys. Conf. Ser.* **2009**, *144*, 12030. [CrossRef]
16. Schriever, U. *Untersuchungen zur Wirkungsweise der Elemente Bor, Titan, Zirkon, Aluminum und Stickstoff in Wasservergüteten, Schweissbaren Baustählen*; Kommission der Europäischen Gemeinschaften, Generaldirek-tion Wissenschaft, Forschung und Entwicklung: Luxemburg, 1991.
17. McMahon, C.J. The role of solute segregation in promoting the hardenability of steel. *Metall. Trans. A* **1980**, *11*, 531–535. [CrossRef]
18. López-Chipres, E.; Mejía, I.; Maldonado, C.; Bedolla-Jacuinde, A.; El-Wahabi, M.; Cabrera, J.M. Hot flow behavior of boron microalloyed steels. *Mater. Sci. Eng. A* **2008**, *480*, 49–55. [CrossRef]
19. Ohmori, Y.; Yamanaka, J. Hardenability of boron-treated low carbon low alloy steels. *Proc. Ann. Symp. Comput. Appl. Med. Care* **1980**, 44–60.
20. Shigesato, G.; Fujishiro, T.; Hara, T. Boron segregation to austenite grain boundary in low alloy steel measured by aberration corrected STEM-EELS. *Mater. Sci. Eng. A* **2012**, *556*, 358–365. [CrossRef]
21. Abenojar, J.; Martinez, M.A.; Velasco, F. Effect of the boron content in the aluminum/boron composite. *J. Alloys Compd.* **2006**, *422*, 67–72. [CrossRef]

22. Nestler, D.; Siebeck, S.; Podlesak, H.; Wagner, S.; Hockauf, M.; Wielage, B. Powder metallurgy of particle-reinforced aluminum matrix composites (AMC) by means of high-energy ball milling. In *Integrated Systems, Design and Technology 2010*; Fathi, M., Holland, A., Ansari, F., Weber, C., Eds.; Springer: Berlin/Heidelberg, Germany, 2011; pp. 93–107.

23. Nestler, D.; Siebeck, S.; Podlesak, H.; Wielage, B.; Wagner, S.; Hockauf, M. Influence of process control agent (PCA) and atmosphere during high-energy ball milling for the production of particle-reinforced aluminum matrix composites. *Materialwiss. Werkstofftech.* **2011**, *42*, 580–584. [CrossRef]

24. Podlesak, H.; Siebeck, S.; Mücklich, S.; Hockauf, M.; Meyer, L.; Wielage, B.; Weber, D. Powder metallurgical fabrication of SiC and Al_2O_3 reinforced Al-Cu alloys. *Materialwiss. Werkstofftech.* **2009**, *40*, 500–505. [CrossRef]

25. Nakashima, S.; Harima, H. Raman investigation of SiC polytypes. *Phys. Stat. Sol.* **1997**, *162*, 39–64. [CrossRef]

26. Meyer, F.D. Festkörperchemische Untersuchungen von Ternären Aluminumcarbiden mit Bor, Silicium und Stickstoff. Doctoral thesis, Albert-Ludwigs-Universität Freiburg, Freiburg, Germany, 1998.

27. Meyer, F.D.; Hillebrecht, H. Synthesis and crystal structure of Al_3BC, the first boridecarbide of aluminum. *J. Alloys Compd.* **1997**, *252*, 98–102. [CrossRef]

28. Madelung, O.; Kück, S.; Werheit, H. *Non-Tetrahedrally Bounded Binary Compounds II, New Series*; Springer: Berlin/Heidelberg, Germany, 2000.

29. Xiang, H.; Li, F.; Li, J.; Wang, J.; Wang, X.; Wang, J.; Zhou, Y. Raman spectrometry study of phase stability and phonon anharmonicity of Al_3BC_3 at elevated temperatures and high pressures. *J. Appl. Phys.* **2011**, *110*, 113504. [CrossRef]

30. Orlovskaya, N.; Lugovy, M. *Boron Rich Solids. Sensors, Ultra High Temperature Ceramics, Thermoelectrics, Armor*; Springer: Dordrecht, The Netherlands, 2011.

metals

MDPI

Article

Roller Burnishing of Particle Reinforced Aluminium Matrix Composites

Andreas Nestler and Andreas Schubert *

Professorship Micromanufacturing Technology, Chemnitz University of Technology, Reichenhainer Str. 70, 09126 Chemnitz, Germany; mft@tu-chemnitz.de
* Correspondence: andreas.schubert@mb.tu-chemnitz.de; Tel.: +49-371-531-34580

Received: 15 December 2017; Accepted: 22 January 2018; Published: 27 January 2018

Abstract: Energy and resource efficient systems often demand the use of light-weight materials with a specific combination of properties. However, these requirements usually cannot be achieved with homogeneous materials. Consequently, composites enabling tailored properties gain more and more importance. A special kind of these materials is aluminium matrix composites (AMCs), which offer elevated strength and wear resistance in comparison to the matrix alloy. However, machining of these materials involves high tool wear and surface imperfections. An approach to producing high-quality surfaces consists in roller burnishing of AMCs. Furthermore, such forming technologies allow for the generation of strong compressive residual stresses. The investigations address the surface properties in the roller burnishing of AMCs by applying different contact forces and feeds. For the experiments, specimens of the alloy AA2124 reinforced with 25% volume proportion of SiC particles are used. Because of the high hardness of the ceramic particles, roller bodies were manufactured from cemented carbide. The results show that roller burnishing enables the generation of smooth surfaces with strong compressive residual stresses in the matrix alloy. The lowest surface roughness values are achieved with the smallest feed (0.05 mm) and the highest contact force (750 N) tested. Such surfaces are supposed to be beneficial for components exposed to dynamic loads.

Keywords: aluminium; aluminium matrix composite; burnishing; finishing; forming; metal forming; metal matrix composite; residual stress; roller burnishing; surface integrity

---

1. Introduction

Aluminium matrix composites are lightweight materials consisting of an aluminium alloy and at least one reinforcing component. Depending on the type of reinforcement they can be classified into:

- Continuous fibre-reinforced AMCs.
- Whisker-reinforced or short fibre-reinforced AMCs.
- Particle-reinforced AMCs.
- Hybrid AMCs with different types of reinforcement.

The type, proportion, and kind of reinforcements significantly influence the properties of such composites. In the majority of cases, an increase of the strength, the Young's modulus, and the wear resistance is aimed for. Consequently, most of AMCs are particle reinforced. Typical particle reinforcements are silicon carbide and aluminium oxide, but titanium diboride, boron carbide, titanium aluminide, or titanium dioxide can also be used [1]. The fabrication is realised by casting processes (e.g., stir casting, squeeze casting, in-situ-casting, spray casting, infiltration), powder metallurgy routes, friction stir processing [2], or selective laser melting [3]. Aluminium matrix composites exhibit a high potential for tribological applications, for example brake discs, brake drums, and cylinder working surfaces of combustion engines. However, there are abrasive and adhesive wear mechanisms depending on the applied load and the properties of the tribological system. An increase of the particle proportion can significantly reduce the wear rate [4]. This can be referred to the behaviour of the ceramic particles acting as load-bearing components. Furthermore, the contact area between the AMC part and the counterpart is markedly reduced [2]. However, the surface structure of the friction partners strongly affects the mechanisms of action in tribological systems.

In general, the surface properties of components influence the functional behaviour of technical systems significantly. In addition to the material, the manufacturing processes are a key factor for the surface integrity. Especially, a mechanical surface modification exhibits a very high potential for a customised improvement of the performance, resulting from changes in the surface structure and the surface layer. There are three basic effects that are involved in mechanical surface modification processes [5]:

- A plastic deformation of the roughness peaks can result in a smoothing effect.
- The forces acting parallel and perpendicular to the surface lead, in combination with a plastic deformation of the surface, to a stretched area with maximum strains at the surface.
- A Hertzian pressure yields to a maximum deformation below the surface.

These effects often involve a decrease of the surface roughness values, work hardening, strong compressive residual stresses, and a grain refinement in the surface layer. Consequently, appropriate mechanical modifications of the surface contribute to an enhancement of the fatigue properties, corrosion resistance, tribological behaviour, and wear resistance.

However, there are many different technologies for a mechanical surface modification, which can be subdivided in burnishing, shot peening, and machine hammer peening processes. The surface properties depend strongly on the specific process parameters and can be varied in a large range.

AlMangour and Yang [6] investigated the surface quality and the mechanical properties after shot peening of stainless steel. The results showed a significant reduction of the surface roughness values. Furthermore, the properties of the surface layer could be modified markedly. The crystalline grain size was approximately halved, but the micro-strain and the absolute values of the compressive residual stresses could be increased considerably. The modified surface properties led to a reduction of the wear volume loss in tribomechanical tests.

However, Scheel et al. [7] studied the influence of different mechanical processes on a modification of the surface. For high strength aluminium alloys, a burnishing process led to stronger and deeper residual stresses compared to shot peening. Consequently, specimens modified by a burnishing process performed better in high cycle fatigue tests in comparison to untreated and shot peened specimens respectively.

For a surface modification of rotationally symmetric components, burnishing processes are preferred because of the simple integration in lathes. Many research studies deal with the burnishing of different alloys, but there is only little information about burnishing of aluminium materials.

El-Axir et al. [8] studied the ball burnishing of the alloy 2014, using a hardened steel ball with a diameter of 8 mm. The surface roughness values could be reduced markedly. For a higher burnishing speed and a burnishing feed in the range of 0.15 mm to 0.25 mm, the smallest surface roughness values were achieved. Moreover, for low speeds multiple burnishing passes should be applied, yet for higher speeds surface deterioration occurs after repeated burnishing.

El-Tayeb et al. [9] investigated the ball burnishing of the alloy 6061 with hardened steel balls. The results show an important influence of the burnishing force and the burnishing speed on the surface roughness. For optimal parameters, an arithmetic mean surface roughness *Ra* of 0.09 μm was achieved.

Nestler and Schubert [10] studied the influence of the machining parameters in diamond smoothing of aluminium matrix composites (AMCs). This process is characterised by a sliding relative movement of the diamond body and the workpiece instead of a primarily rolling relative movement of the roller body and the workpiece. The surface roughness values could be reduced markedly by diamond smoothing, especially for the smallest feed applied. In this case, mean surface roughness values after smoothing were about 0.04 µm for *Ra* and 0.7 µm for *Rz* using a diamond body with a spherical radius of 2 mm. Furthermore, residual stresses in the matrix alloy of about −400 MPa could be gained. The speed varied did not show any significant influence on the surface properties.

However, there are no studies in roller burnishing of heterogeneous materials like aluminium matrix composites, although they gain more and more in importance. Consequently, appropriate parameters for roller burnishing of particle-reinforced AMCs have to be found. The focus of the research is on the surface structure and the residual stress state. Both aspects can influence the functional behaviour, especially fatigue properties or corrosion resistance, markedly.

2. Materials and Methods

For the experiments an aluminium matrix composite consisting of the alloy AA2124 and SiC particles with a volume proportion of 25% was used. The AMC is produced by a powder metallurgy route comprising a high energy mixing process and subsequent hot isostatic pressing for powder consolidation. Afterwards, the billets are extruded to bars applying an extrusion ratio of about 50:1. For an increase of the material strength, the bars are heat treated to the condition T4 (solution annealed, quenched and naturally aged). Figure 1 shows the microstructure of the material in the longitudinal direction.

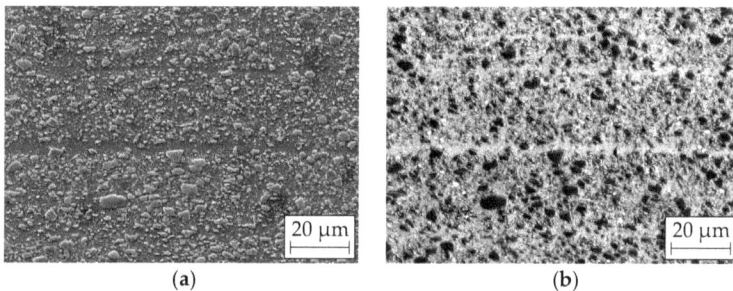

(a)   (b)

Figure 1. SEM (scanning electron microscope) micrographs of the microstructure of the aluminium matrix composite in the longitudinal direction: (**a**) SE (secondary electrons) mode; (**b**) QBSD (quadrant backscatter electron detector).

The mean particle size lies in the range of 2 µm to 3 µm. However, there are many smaller and some larger particles, too. The cross-section polish for the longitudinal direction reveals a slight banding, resulting from the extrusion process. Table 1 represents the mechanical properties of the material identified for the longitudinal direction (hardness excluded).

Table 1. Mechanical properties of the aluminium matrix composite with standard deviations out of three tests (for the hardness out of five tests).

Parameter	Characteristic Value
Yield strength $R_{p0.2}$	513 MPa ± 5 MPa
Ultimate tensile strength R_m	699 MPa ± 8 MPa
Young's modulus E	117 GPa
Elongation without reduction of area A_g	4.6% ± 1.0%
Vickers hardness HV	219 HV10 ± 14 HV10

The finally premachined specimens for the investigations in roller burnishing exhibit a diameter of 23 mm and a length of 20 mm. Both faces are chamfered with an angle of 45° and a size of 1 mm. For clamping, one face of each specimen comprises a blind hole with a diameter of 8 mm, and on the opposite side there is a small centre hole. The values for the surface roughness depth Rz before roller burnishing were in the range of about 5 µm to 7 µm.

The experiments were carried out on a precision lathe of the type SPINNER PD 32 (SPINNER Werkzeugmaschinen GmbH, Sauerlach, Germany). A clamping of the specimens with a mandrel allowed machining the complete cylindrical length. Because of the high forces in roller burnishing, for this process clamping was supported by an additional live centre on the opposite side. The turning tools for premachining were carried by the disc turret (BARUFFALDI S.p.a., Tribiano (MI), Italy). Because of the highly abrasive effect of the hard ceramic particles CVD (chemical vapour deposition), diamond tipped indexable inserts were used. The turning tools exhibit a polished rake face and a very sharp cutting edge with a rounding of about 3 µm. For final premachining, an insert of the type VCGW 110304 (DTS GmbH – Diamond Tooling Systems, Kaiserslautern, Germany) screwed on a tool holder of the kind SVVCN 1212 F11 (WNT Deutschland GmbH, Kempten, Germany) was applied. Final premachining was realised with a feed of $f = 0.14$ mm, a cutting speed of $v_c = 150$ m/min, and a depth of cut of $a_p = 0.25$ mm.

For burnishing, a discoid roller body with a radius of 21 mm in the rolling direction and a radius of 2 mm perpendicular to the rolling direction was used. The roller body consists of cemented carbide to withstand the hard ceramic particles in the AMC. The burnishing tool (BAUBLIES AG, Renningen-Malmsheim, Germany) was mounted on a three-axis force dynamometer of the type Kistler 9257A (Kistler Instrumente AG, Winterthur, Switzerland) to monitor the rolling force. This force can be varied by a change of the helical compression spring, its pretension, and the infeed. The tool was adjusted to an angle of 90° between the direction of the burnishing force and the feed direction, which corresponds to the axial direction of the specimens.

For all experiments, the rolling speed was kept constant at 150 m/min. The final rolling force was adjusted by the infeed amounting to about 0.2 mm. The feed was varied in the range of 0.05 mm to 0.15 mm and the rolling force from 250 N to 750 N using a full factorial design for these both parameters. For each combination of parameters tested, three specimens were machined to assess the process stability. The process parameters for roller burnishing are presented in Table 2. An emulsion cooling with a concentration of approximately 5% was used to reduce the tendency for adhesion of the aluminium based material on the indexable inserts and the roller body.

Table 2. Parameters for roller burnishing.

Process Parameter	Values
Rolling force F	250 N; 500 N; 750 N
Rolling feed f	0.05 mm; 0.1 mm; 0.15 mm
Rolling speed v	150 m/min

The surface roughness in the axial direction was measured using a stylus instrument of the type Mahr LD 120 (Mahr GmbH, Göttingen, Germany). The measuring length is 4 mm and filtering of the profile is done in accordance to ISO 11562. Because of the comparatively strong influence of surface imperfections on the roughness values, each specimen was measured thrice at different positions. Additionally, for each parameter combination tested, one three-dimensional surface profile was generated with the same measurement equipment including a supplemental cross table. For these measurements, the distance of the measuring points is 1 µm for the circumferential and the axial direction. The measuring field has a size of 2 mm in the axial direction and 0.5 mm in the circumferential direction. For a detailed examination of the surfaces, the 3D profiles were leveled by the subtraction method. Afterwards, a form removal was done by applying a fifth degree polynomial filter (Digital Surf, Besançon, France). These data were used for detailed 3D images of the surface structure, representing a smaller area than measured. A characterisation of the porosity of the surfaces generated by roller burnishing requires further mathematical procedures. After using a robust Gaussian filter with a cut-off wavelength of 0.08 mm, a line by line leveling followed to remove the kinematic roughness.

Furthermore, SEM (scanning electron microscope) micrographs (Carl Zeiss AG, Oberkochen, Germany) of the specimens' surfaces were obtained using a Zeiss LEO 1455VP microscope to characterise the surface structure and the imperfections. The residual stresses in the surface layer were determined by X-ray diffraction analysis performed with a Siemens D5000 diffractometer (Siemens Aktiengesellschaft, Munich, Germany). The measurements were done with a cobalt anode in the lattice planes {420} of the aluminium alloy using $\sin^2 \psi$ method. Thereby, an area with a diameter of about 2 mm was detected.

3. Results and Discussion

3.1. Influence of the Feed and the Rolling Force on the Surface Structure

A detailed understanding of the rolling process and a profound discussion of the results require ample information about the structure of the premachined surface. Figure 2 presents details of the surface structure of a premachined specimen.

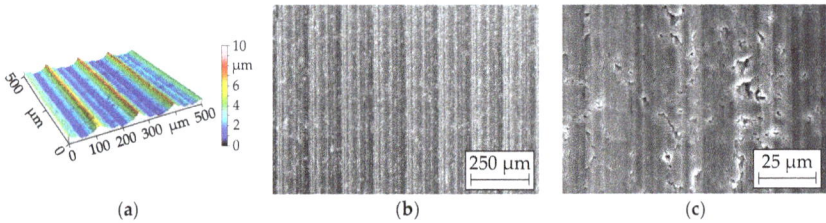

Figure 2. Surface structure of a premachined specimen (a_p = 0.25 mm, f = 0.14 mm, v_c = 150 m/min): (**a**) 3D surface profile; (**b**) SEM micrograph (SEM magnification 100); (**c**) SEM micrograph (SEM magnification 1000).

The three-dimensional surface profile shows distinct feed marks with a distance to each other, corresponding to the turning feed of 0.14 mm. Furthermore, form deviations of the arc-shaped tool corner and cutting edge chipping are reflected in this profile. A SEM micrograph with an overview section (Figure 2b) confirms the feed marks. Additionally, numerous voids with different sizes and shapes are revealed by a SEM micrograph with a larger magnification (Figure 2c).

The feed in roller burnishing has a significant influence on the surface structure. Figure 3 represents 3D surface profiles for a rolling force of 500 N.

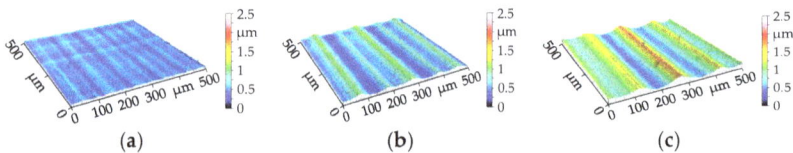

Figure 3. Influence of the rolling feed on the surface structure (F = 500 N): (**a**) f = 0.05 mm; (**b**) f = 0.1 mm; (**c**) f = 0.15 mm.

Surfaces generated with a rolling feed of 0.05 mm are comparatively smooth and do not exhibit a distinct "waviness". This can be explained by the very small theoretical roughness, represented in Figures 4 and 5. For an increase of the feed, a kinematic roughness is regularly formed on the surface. The distance of the "single wave structures" complies with the rolling feed. However, there is an alternation of a higher and a lower "wave". This may be due to small form deviations of the roller body. For the geometrical dimensions applied, one revolution of the roller body requires about two revolutions of the specimen, which underpins this assumption.

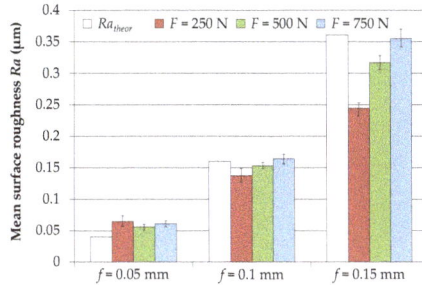

Figure 4. Influence of the feed and the rolling force on the arithmetic mean surface roughness.

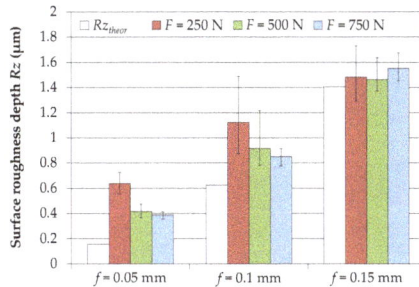

Figure 5. Influence of the feed and the rolling force on the surface roughness depth.

The most common parameters for the characterisation of the surface structure are the arithmetic mean surface roughness *Ra* and the surface roughness depth *Rz*. Figure 4 shows the influence of the feed and the rolling force on the arithmetic mean surface roughness.

The error bars represent the scattering of the values starting at the lowest values and ending at the highest value. Each coloured bar incorporates nine measurements (three measurements on each specimen at different locations). The white bars indicate the calculated values using the equation:

$$Ra_{theor} \approx \frac{f^2}{31.2 \cdot r} \tag{1}$$

The variable *r* stands for the radius of the roller body measured in the feed direction. For the smallest feed applied, the measured values are in the same range, but higher than the calculated value. This can be attributed to minor surface imperfections, especially voids and grooves. However, an increase of the feed results suggests the increasing influence of the rolling force. For lower forces, the measured values are smaller than the theoretical values. This is due to the stress state in roller burnishing, characterised by a high hydrostatic proportion. However, only deviatoric stresses contribute to the forming process. For higher feeds, the deviatoric stresses were not sufficient to form the theoretical roughness profile completely. It can be concluded that an increase of the rolling force involves a better forming of the theoretical roughness profile and consequently higher values for the arithmetic mean surface roughness, especially for higher feeds.

However, there are some differences for the surface roughness depth, illustrated in Figure 5.

The determination of the error bars is in accordance with Figure 4. For a feed of 0.05 mm, the measured values are significantly higher than the theoretical value calculated with the equation

$$Rz_{theor} \approx \frac{f^2}{8 \cdot r} \tag{2}$$

This can be ascribed to surface imperfections, which have a stronger influence on the surface roughness depth than on the Ra-values. For the smallest force, the roughness profiles reveal deep voids increasing the roughness values. With an increase of the force the number and the depth of the voids decrease, resulting in smaller mean values and lower scattering. A reduction of the surface porosity requires strong compressive stresses in the stress deviator. This is supported by an enlargement of the rolling force. For an increase of the feed, the formation of the kinematic roughness profile gains in importance compared to the surface imperfections. This leads, for a feed of 0.15 mm, to measured surface roughness values Rz in the range of the theoretical value. Consequently, there is an increase of the surface roughness depth with a raising feed, which complies with the Ra-values. The surface roughness values for Ra and Rz lie in the same range like the values after diamond smoothing of AMCs [10].

The valley void volume is used as a measure for the porosity of the surfaces generated by roller burnishing. Figure 6 shows the influence of the feed and the rolling force on the valley void volume.

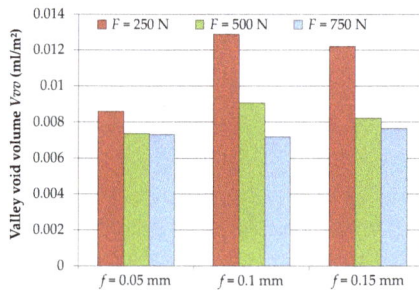

Figure 6. Influence of the feed and the rolling force on the valley void volume.

There is no significant effect of the feed on the valley void volume except for the lowest force applied. For all feeds tested, the rolling force of 250 N results in higher values for the valley void volume. This indicates that the smallest force is not sufficient for a prevention of the formation of voids or its closure. It is evidenced in particular for the feeds 0.1 mm and 0.15 mm, resulting in the highest values. However, an increase of the rolling force leads to a reduction of the valley void volume. This is confirmed by SEM micrographs (Figure 7) and can be referred to stronger compressive stresses in the stress deviator.

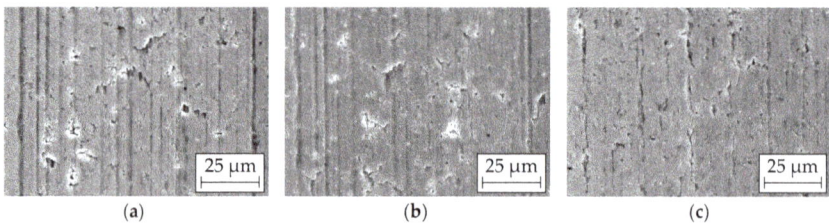

Figure 7. SEM micrographs of details of the surfaces generated with f = 0.15 mm: (**a**) F = 250 N; (**b**) F = 500 N; (**c**) F = 750 N.

Surfaces generated with a rolling force of 250 N exhibit comparatively large and deep voids (Figure 7a). Furthermore, slight grinding grooves of the roller body are transferred to the surface of the specimens running in the circumferential direction. For an increase of the rolling force, the compressive stresses in the stress deviator become stronger. Consequently, the area and the depth of the voids decrease.

3.2. Influence of Roller Burnishing on the Residual Stresses

An evaluation of the residual stresses after roller burnishing requires information about the residual stress state of the premachined specimens. The X-ray diffraction analysis of a premachined specimen showed residual stresses of about −32 MPa in the axial direction and −95 MPa in the circumferential direction, both for the aluminium alloy. The roller burnishing process generally resulted in a significant increase of the absolute values of the residual stresses. However, for the feeds applied only small differences concerning the residual stress state of the burnished surface occurred. Figure 8 represents the results for a feed of 0.15 mm.

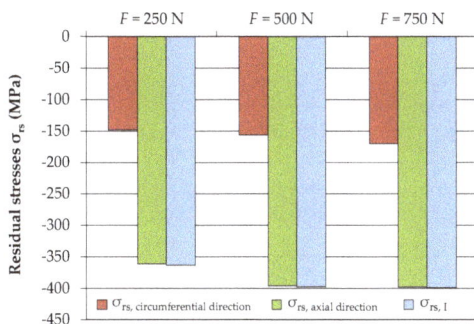

Figure 8. Influence of the rolling force on the residual stresses for a feed of 0.15 mm.

The red bars represent the residual stresses in the circumferential direction and the green bars the residual stresses in the axial direction. The first principle residual stresses (blue bars) approximately comply with the residual stresses in the axial direction. Consequently, the first principle axis has nearly the same direction as the rotational axis of the specimens. This can be attributed to the strongest compression of the surface area in the axial direction, resulting from the kinematic roughness of the premachined specimens. The diagram shows, for a feed of 0.15 mm and an increasing force, a slight growth of the absolute values of the residual stresses. This may be due to a decrease of the porosity of the surfaces, characterised by the valley void volume. The absolute values of the residual stresses in the axial direction correspond nearly to the yield strength of the matrix alloy and are significantly higher than the absolute values of the residual stresses in the circumferential direction. The reason for this is the stronger reduction of the surface size of the premachined specimens, exhibiting a kinematic roughness in the axial direction, to a nearly smooth cylindrical surface. The residual stresses in the axial direction correspond to the values measured on the surface of burnished AA2024-T351 [7].

4. Conclusions

Based on experimental investigations in roller burnishing of particle-reinforced aluminium matrix composites and detailed surface analyses the subsequent conclusions can be drawn.

- Roller burnishing of aluminium matrix composites leads to a significant decrease of the surface roughness values. The lowest surface roughness value for *Ra* is about 0.05 μm.
- For a feed larger than 0.05 mm, the surface structure exhibits a kinematic roughness corresponding to the calculated roughness values.

- The roller burnishing process results in strong compressive residual stresses in the aluminium matrix. The absolute values of the residual stresses in the axial direction are much higher than the absolute values of the residual stresses in the circumferential direction.
- The results in roller burnishing concerning roughness and residual stress state at the surface are comparable to the findings for diamond smoothing.
- The modification of the surface properties is expected to improve the wear and the fatigue behaviour.
- Further investigations have to be done to characterise the changes in the microstructure and to analyse the residual stresses in the subsurface area.

Acknowledgments: The authors gratefully acknowledge funding by the German Research Foundation (Deutsche Forschungsgemeinschaft, DFG) within the framework of the Collaborative Research Center SFB 692.

Author Contributions: Andreas Nestler and Andreas Schubert conceived and designed the experiments; Andreas Nestler performed the experiments, analysed the data, and wrote the paper.

Conflicts of Interest: The authors declare no conflict of interest.

References

1. Ranjith, R.; Giridharan, P.K.; Kumar, G.S.; Seenivasan, N. A review on advancements in aluminium matrix composites. *Int. J. Adv. Eng. Techol.* **2016**, *VII*, 173–176.
2. Shojaeefard, M.H.; Akbari, M.; Asadi, P.; Khalkhali, A. The effect of reinforcement type on the microstructure, mechanical properties, and wear resistance of A356 matrix composites produced by FSP. *Int. J. Adv. Manuf. Technol.* **2017**, *91*, 1391–1407. [CrossRef]
3. AlMangour, B.; Grzesiak, D.; Yang, J.-M. Selective laser melting of TiC reinforced 316L stainless steel matrix nanocomposites: Influence of starting TiC particle size and volume content. *Mater. Des.* **2016**, *104*, 141–151. [CrossRef]
4. Prasat, S.V.; Ram, A.S.; Sivaprakash, S.; Suresh, K. A review on mechanical and wear behaviour of aluminium metal matrix composites. *Int. Res. J. Eng. Technol.* **2017**, *4*, 1261–1265.
5. Schulze, V.; Bleicher, F.; Groche, P.; Gou, Y.B.; Pyun, Y.S. Surface modification by machine hammer peening and burnishing. *Cirp. Ann. Manuf. Technol.* **2016**, *65*, 809–832. [CrossRef]
6. AlMangour, B.; Yang, J.-M. Improving the surface quality and mechanical properties by shot peening of 17-4 stainless steel fabricated by additive manufacturing. *Mater. Des.* **2016**, *110*, 914–924. [CrossRef]
7. Scheel, J.; Prevéy, P.; Hornbach, D. Safe Life Conversion of Aircraft Aluminum Structures via Low Plasticity Burnishing for Mitigation of Corrosion Related Failures. In Proceedings of the Department of Defense Corrosion Conference, Washington, DC, USA, 10–14 August 2009.
8. El-Axir, M.H.; Othman, O.M.; Abodiena, A.M. Study on the inner surface finishing of aluminum alloy 2014 by ball burnishing process. *J. Mater. Process. Technol.* **2008**, *202*, 435–442. [CrossRef]
9. El-Tayeb, N.S.M.; Low, K.O.; Brevern, P.V. On the surface and tribological characteristics of burnishes cylindrical Al-6061. *Tribol. Int.* **2009**, *42*, 320–326. [CrossRef]
10. Nestler, A.; Schubert, A. Effect of machining parameters on surface properties in slide diamond burnishing of aluminium matrix composites. *Mater. Today Proc.* **2015**, *2S*, S156–S161. [CrossRef]

![metals logo] *metals*

MDPI

Article

Particle-Reinforced Aluminum Matrix Composites (AMCs)—Selected Results of an Integrated Technology, User, and Market Analysis and Forecast

Anja Schmidt [1,*], Steve Siebeck [2], Uwe Götze [1], Guntram Wagner [2] and Daisy Nestler [3]

[1] Management Accounting and Control, Faculty of Economics and Business Administration, Chemnitz University of Technology, 09107 Chemnitz, Germany; uwe.goetze@wirtschaft.tu-chemnitz.de
[2] Composites and Material Compounds, Faculty of Mechanical Engineering, Chemnitz University of Technology, 09107 Chemnitz, Germany; steve.siebeck@mb.tu-chemnitz.de (S.S.); guntram.wagner@mb.tu-chemnitz.de (G.W.)
[3] Textile Plastic Composites and Hybrid Compounds, Faculty of Mechanical Engineering, Chemnitz University of Technology, 09107 Chemnitz, Germany; daisy.nestler@mb.tu-chemnitz.de
* Correspondence: anja.schmidt@wirtschaft.tu-chemnitz.de; Tel.: +49-(0)371-531-34172

Received: 31 December 2017; Accepted: 19 February 2018; Published: 22 February 2018

Abstract: The research and development of new materials such as particle-reinforced aluminum matrix composites (AMCs) will only result in a successful innovation if these materials show significant advantages not only from a technological, but also from an economic point of view. Against this background, in the Collaborative Research Center SFB 692, the concept of an integrated technology, user, and market analysis and forecast has been developed as a means for assessing the technological and commercial potential of new materials in early life cycle stages. After briefly describing this concept, it is applied to AMCs and the potential field of manufacturing aircraft components. Results show not only technological advances, but also considerable economic potential—the latter one primarily resulting from the possible weight reduction being enabled by the increased yield strength of the new material.

Keywords: aluminum matrix composites; light-weight materials; aircraft industry; integrated technology; user; and market analysis and forecast; cost and revenues

1. Introduction

By the reinforcement of aluminum materials, improved mechanical properties can be achieved [1,2], which is promising for applications, e.g., in automotive or aircraft industries. However, corresponding research and development activities are in an early stage, the transferability of their results in industrial applications is uncertain, and high risks exist. Usually it takes a long time to introduce new materials into the automotive and the aircraft industry. Against this background, it is important to appraise the technological, as well as commercial, potential [3] of the material innovation as early as possible. For appraisal, the methodology of an integrated technology, user, and market analysis and forecast is suggested.

Firstly, this paper presents the basic structure of the methodology which has been explored and elaborated in the Collaborative Research Center SFB 692 [4,5]. Secondly, selected results of the technology analysis and forecast, as well as user and market analysis and forecast of powder-metallurgically produced particle-reinforced aluminum matrix composites (AMCs) [6–10] are outlined and reflected. Finally, conclusions are drawn. The results will give some evidence about the technological, as well as commercial, potential of AMCs, can be used for directing research activities, and—in case a sufficient technology maturity can be achieved—might contribute to their dissemination on the markets.

2. Methodology

For the appraisal of the commercial potential of a new technology, profound knowledge about the potentials and disadvantages of this technology, the requirements and other characteristics of potential users, and the market seems to be necessary. In order to contribute to the enhancement of instruments of a systematic and effective innovation control, in the SFB 692 such a methodology of an integrated technology, user, and market analysis and forecast has been explored and elaborated [4,5]. Technology analysis and forecast comprises the description and classification of the technology as well as the view on competing technologies. In this paper, properties of powder-metallurgically produced AMCs are described and compared to a reference matrix material without any reinforcements and a commercially-available AMC (Duralcan) produced by means of a melt-metallurgical production method. User analysis and forecast deals with the identification of potential application fields and potential users in different stages of the value chain, the requirements and demands of these users, as well as their willingness to pay. Therefore, it has to comprise an analysis of innovation-dependent costs and benefits which are relevant from the perspective of users. For this task, life cycle costing as well as (other) instruments for the cost (and revenue) appraisal of material, product, and process technologies are available [11]. Since a commercialization can only succeed if the new technology fits the requirements and demands, this has to be checked particularly. Market analysis and forecast is intended to characterize the market and competitors and to forecast their performance. Finally, technology appraisal integrates the results of analyses and forecasts to an overall appraisal of the economic potential of a new technology. Figure 1 shows the basic structure of the methodology.

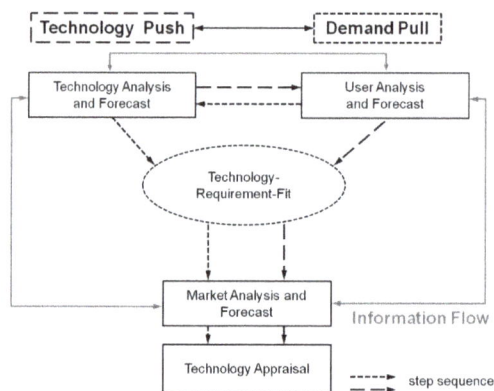

Figure 1. Basic structure of an integrated technology, user, and market analysis and forecast [4].

The arrangement of the several analysis and forecast components depends on the fact if innovations are pushed onto the market (technology push) or if they are developed because of concrete demand of users (demand pull). In case of a technology push (as it is intended in the case of powder-metallurgically produced particle-reinforced aluminum matrix composites) it has to be questioned initially what the new (material and process) technology can accomplish, which benefit can be derived, and who can benefit. Then it has to be clarified which users (characterized by specific requirements, demands, and willingness to pay) could apply the new technology in which stage of the value chain. If the technology fits the requirements, a market analysis can be realized within a third step, followed by an appraisal of the commercial potential of the technology. Sections 3–5 show selected results and more methodical details of the integrated technology, user, and market analysis and forecast applied to powder-metallurgically produced AMCs.

3. Aluminum Matrix Composites Produced by Means of Mechanical Alloying—Technology Analysis and Forecast

Aluminum materials are reinforced in order to improve the mechanical properties, such as the Young's modulus, the tensile strength, the yield strength, and the wear resistance [1,12]. The intensity of the individual improvement on the properties due to the reinforcement depends heavily on the type, size, amount, and distribution of the reinforcement particles. To achieve the desired properties, a high degree of dispersion, complete embedding of the particles within the metal matrix, and the development of a suitable interface are required [2,13]. In general, the smaller the particles, the higher the possible property improvement, but the more difficult the manufacturing becomes.

The production of particle-reinforced aluminum matrix composites (AMC) can be realized by two different ways, the powder-metallurgical and the melt-metallurgical processing. The first presents significant advantages regarding material properties. The latter is not suited to reach a high dispersion degree of nano-scaled particles in a metal matrix. Furthermore, the small particles would react with the melt and disappear or form undesirable phases [14,15]. Therefore, powder-metallurgical techniques are focused in the SFB 692 and in this paper. For this purpose, the methods high-energy ball milling, hot isostatic pressing, and warm extrusion are used. First step is the manufacturing of the composite powders (metallic powder dispersed with particles) by means of high-energy ball milling (HEBM). It provides a homogeneous distribution of the reinforcement particles. The formation of the final composite powder goes through several stages, which is typical for ductile-brittle powder systems (Figure 2). Due to the high ductility of the Al-powder, the first stage of the milling process is characterized by the formation of deformed flat Al-particles with a simultaneous attachment of the SiC-reinforcement to the surface of these flakes. In the next stage, the cold welding of the flakes amongst themselves dominates and leads to the production of large composite particles with lamellar structure (mixture of alternating reinforced and unreinforced lamellae). Further milling of the lamellae causes an increase in mixing and, thus, an improvement of the degree of dispersion.

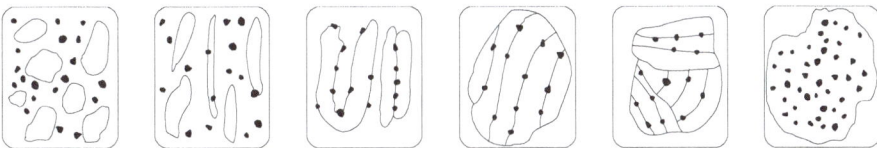

Figure 2. Schematic formation of composite powder during high energy ball milling (based on [16]).

The composite powders have to be compacted in a subsequent step. A compaction method which leads to fully dense materials is the hot-isostatic pressing. It is able to transfer the composite powders into semi-finished products. Due to high scalability, the production on an industrial scale is easily possible.

The data used within this article has been determined as part of the Collaborative Research Center SFB 692. The production of the materials and their analysis was carried out under the conditions outlined below.

The aluminum alloy that was used as matrix material was supplied in the form of a commercial, gas-atomized, spherical powder with a particle size fraction <100 µm. The chemical composition (in weight-percent) of the alloy was about 4.1% Cu, 0.7% Mg, 0.8% Mn, 0.1% Si, 0.2% Fe, balance Al. Fine-grained SiC alpha phase powder with a fraction of about 1 µm, as well as a nano-sized beta phase with a fraction size smaller than 200 nm were used as reinforcing components. The preparation of the AMCs was carried out for the three volume fractions 5, 10, and 15 percent by volume.

The composite powder was processed in a high-energy ball mill Simoloyer® CM08 from Zoz Company (Wenden, Germany). Milling was performed for at least four hours in air atmosphere. Details on the preparation of this kind of AMCs are already published in [6–10].

Compaction for all materials was then performed by hot isostatic pressing at 450 °C for 3 h and at a pressure of 1100 bar. Finally, the material was extruded in a temperature range between 355 and 370 °C to produce semi-finished square bars with a cross-section of 15 × 15 mm. The extrusion was performed with a punch speed of 2 mm/s and an extrusion ratio of 42:1.

For the characterization of strength and ductility, cylindrical tensile specimens (with an aspect ratio of the gauge length of three) were machined from the billets in the direction of extrusion. Quasi-static tensile tests were performed at room temperature in a conventional testing machine (Zwick-Roell) with a constant cross-head speed corresponding to an initial strain rate of 10^{-3} s^{-1}. At least three tests were performed for the different material conditions, in particular to provide better statistics on fracture strains of the AMCs. Figure 3 shows three AMCs produced in this way. Some related material characteristics are listed in Table 1.

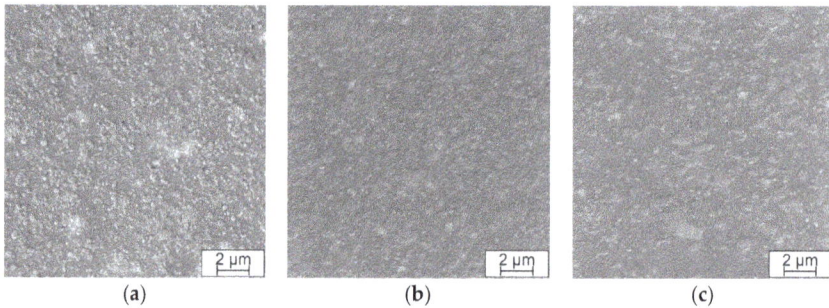

(a)   (b)   (c)

Figure 3. SEM images of AMCs produced by means of high-energy ball milling, hot isostatic pressing and extrusion in the T4 condition; matrix alloy AA2017; reinforced with SiC particles: (**a**) 15 vol % β-phase with size < 0.2 μm; (**b**) 15 vol % α-phase with size < 1 μm; and (**c**) 10 vol % α-phase with size < 1 μm.

Table 1. Selected properties of powder-metallurgically produced AMCs compared to the reference matrix material without any reinforcements and a commercially available AMC (Duralcan) produced by means of a melt-metallurgical production method.

	Reference Material			EN AW-2017 AMC Materials		
	AA 2017	Duralcan	Duralcan	(PVW Production)		
	T4 [1]	F3S20S [2] Cast	F3S20S [2] Extruded	T4 (a)	T4 (b)	T4 (c)
SiC content in vol %	none	20	20	15	15	10
SiC size in μm	-	12	12	0.2	1.0	1.0
Tensile strength in MPa	425	218	355	683	630	580
Yield strength in MPa	275	191	253	540	480	465
Elongation in %	22.0	0.4	2,8	5.5	5.0	12.0
E modulus in GPa	73	99	113	92	100	90
α [3] in 10^{-6} K^{-1}	23.4	17.1	17.1	18.9	17.9	19.5
K [4] in W/m K	141	192	192	123	126	133

[1] [17]; [2] [18]; [3] Thermal expansion coefficient; [4] Thermal conduction.

The authors have also already dealt with the fatigue behavior of AMCs [19,20]. On the one hand, it becomes clear that an improvement is achieved especially at high load amplitudes above 150 MPa. On the other hand, the behavior at stress amplitudes below 140 MPa is critical, since a fatigue limit does not occur. The causes are often larger intermetallic precipitations resulting from contaminations. A particularly high quality of the composites is, therefore, a basic requirement for applications with cyclic loading. Further specific investigations of the material behavior are indispensable for the transfer into the concrete application.

4. User and Market Analysis and Forecast

As a base for the appraisal of the commercial potential of the properties of powder-metallurgically produced AMCs, *user analysis and forecast* has to identify application fields and potential users and has to analyze users' requirements and willingness to pay for the improved properties. For this, market research instruments, expert interviews, database analyses, creative techniques, instruments of requirements management, such as quality function deployment, the lead user-approach, as well as life cycle costing and (other) instruments for the cost (and revenue) appraisal can be used [5]. Potential application fields of AMCs can be derived from those of materials with similar properties captured by material data bases. For example, the aluminum material data sheets edited by the German Institute for Standardization (DIN) highlight a high mechanical strength, a high fatigue strength, as well as (very) good machining characteristics as essential characteristics of the material EN AW 2017A. Derivable typical applications are high-strength structural components in aircraft construction, automotive engineering, or machine construction [5,21]. Since aircraft construction seems to be a quite promising field of application at a first glance, it is focused in the following considerations.

In addition to titanium, steel, and composite materials, aluminum is a major aircraft material. In aircraft construction, aluminum alloys are especially used in structural components of the fuselage and the airfoil wings [22,23]. From the perspective of an aircraft component manufacturer as potential user, relevant requirements and demands of airlines (regarding properties of aircrafts) and aircraft manufacturers (regarding properties of aircraft components) are important to be analyzed and forecast. A high strength, damage tolerance, as well as corrosion and fatigue resistance, a low weight, good machinability, and low costs are requirements to be considered for the materials [5,22,23]. Concrete, specified requirements can be identified, analyzed, and forecast by the cooperation with a potential lead user (e.g., an aircraft component manufacturer). Lead users are users whose needs become general in the future. They are familiar with current and potential future conditions, can provide design data and experience for specifying requirements [24]. Although specified data are not available in this case until now, it can be assumed that the improved (tensile and yield strength, thermal conduction) or at least competitive (E modulus, thermal expansion) material properties of powder-metallurgically produced AMCs compared to the reference materials (Table 1) provide potentials of a better fulfillment of user requirements. Additionally, from the improved yield strength a considerable lightweight construction potential for components made of AMCs can be derived because of reducible material cross-sections. Taking 250 MPa as a reference value of yield strength, 500 MPa (approx. yield strength of powder-metallurgically produced AMCs, Table 1) represents a doubling of the value. Based on a constant component strength, a reduction of the material cross-section by 50%, a corresponding saving of material quantity, as well as weight reduction of 50% are enabled by the usage of powder-metallurgically produced AMCs [5,25]. These potentials might (over-)compensate the disadvantage of a lower elongation. A lower elongation of a material corresponds to a lower reserve of permanent deformation before breakage [5]. The relevance of a lower elongation of the AMC materials depends on the specific application. Achieved improvements of fatigue behavior (Section 3) can also be a potential. However, further investigations on this are necessary. If potentials of improved properties dominate and are perceived by the users, users can be expected to accept higher costs of AMCs and, therefore, higher prices of aircraft components. Section 5 demonstrates calculations of relevant monetary effects.

Analysis and forecast of markets of aluminum alloys and competing materials has to find out which companies act in the markets and how can the competition situation be characterized from the perspective of material manufacturers, aircraft component manufacturers, aircraft manufacturers, and airlines. For this, instruments such as industry analysis, competitor analysis, market structure analysis, techniques for determination of market attractiveness in portfolio analyses, quantitative forecast methods, and, again, instruments of market research are usable [5,26]. A popular instrument for analyzing the competition in an industry is Porter's Five Forces Framework [26]. Here, only some selected, particularly relevant "forces" of the competition are shortly characterized [5]:

- The market potential of the innovative AMC materials can be derived from the potential of the aircraft market. In the year 2016, the biggest aircraft manufacturers Boeing and Airbus delivered in sum 1436 airplanes (1397 in 2015) and achieved sales revenues of €66.6 billion (Airbus) and $94.6 billion (Boeing) [27–30]. Regarding aviation, for example in Germany, a compound annual sales growth rate of 0.9% can be expected for the period 2016–2021 [31]. This implies a large and stable demand of aircrafts, aircraft components, and materials for manufacturing these components. Because of their small number, the *bargaining power of* the aircraft manufacturers, which are potential *buyers* of the components and materials, tends to be high.
- Accordingly, the aircraft industry seems to be attractive market for *potential entrants*. At the same time, *intensity of competitive rivalry* is relatively high and the long development cycles and cost-intensive activities of research and development constitute significant barriers to entry especially for small- and medium-sized companies. Development and introduction of new materials and manufacturing processes into the aircraft industry might need decades. High safety standards have to be considered and are monitored by government agencies. In Europe, for example, the European Aviation Safety Agency (EASA) is responsible for certification of airworthiness of civil aircrafts [32,33].
- Manufacturers of competing or substitute materials or companies which are able to produce and supply AMCs can be identified by means of databases, Internet platforms [34], or classified directories. The *bargaining power* of these companies, as well as *of suppliers* of ingredients might be very different, depending on the specificity of the material, and has to be verified.
- The identification as well as comparative analyses of relevant *substitute materials* is advisable already during the technology analysis and forecast. The findings have to be reflected against the requirements of the users of the materials and should be incorporated in the market analysis and forecast. Major aircraft materials, in addition to aluminum, are, as mentioned, steel, titanium, and composites, such as carbon fiber reinforced plastic, competing with each other, depending on the application. (Polymer matrix) composite materials, such as carbon fiber composites, tend to replace aluminum and other metal materials in the aircraft industry (for example, as the Boeing 787 shows) due to higher strength and stiffness and a lower weight [22,32,35]. However, disadvantages of these composites are especially high material and manufacturing costs, but also further shortcomings such as a lower reparability and recyclability [23,36]. Furthermore, due to achieved weight-reductions and improved material properties some opposite trends such as the usage of aluminum-lithium alloys exist [22,23]. Additionally, metal matrix composites (including particle-reinforced AMCs) are recognized as promising materials [37,38].

Results of technology analysis and forecast (properties of powder-metallurgically produced AMCs) as well as user and market analysis and forecast (especially requirements of high strength and low weight; high market potential in aircraft industries) are a base for technology appraisal.

5. Technology Appraisal

Technology appraisal is intended to evaluate the commercial potential of a new technology such as powder-metallurgically produced AMCs in an early stage. The commercial potential can be understood as the expected extent of achieving sustainable profits (defined as difference of long-term revenues and costs) [3,5,39]. Information regarding the commercial potential can be a base for material and technology selection, for directing research activities towards specific application fields and design alternatives, and for decisions about the continuation or stop of research activities.

Technology appraisal comprises two steps: non-monetary and monetary appraisal of economic effects of development, manufacturing, and usage of a new technology. *Non-monetary appraisal* is applicable for preselection of promising technologies which have to be analyzed in greater detail and monetarily evaluated in the second step. Considering this, efforts of data collection can be limited. For non-monetary appraisal, criteria of the *resource-based view* of strategic management should be considered. According to this approach, resources have to be valuable, rare, costly to imitate, as well as not substitutable, and exploited by the organization for achieving sustained competitive advantages and above-average profits in markets [5,40,41]. Powder-metallurgically produced AMCs basically seem to be suitable for achieving these advantages. The results of the technology analysis (Section 3) show considerable assets of the powder-metallurgically produced AMCs compared to the reference materials. In combination with further properties (e.g., machinability [42], low or moderate costs, recyclability), the material potentially features a certain uniqueness and *rarity*, and a specific *value* for users might be gained from this. The powder-metallurgically produced AMCs are *costly to imitate* by competitors of a company when the explored knowledge regarding the materials and their processing is specific (which is supposed here). Furthermore, it seems to be conceivable that the imitability of the knowledge can be restricted by patents or exclusive contracts with users. However, the market characterizations (Section 4) outline a *substitutability* of materials in the aircraft industry. Whether powder-metallurgically produced AMCs could be substituted or not strongly depends on the further developments in materials research regarding AMCs themselves, as well as competing materials, such as carbon fiber composites. Currently, AMCs are investigated by institutes and companies and there are some positive trends of their use [37,38,43]. Finally, a company's ability to exploit the resources (the AMC materials as well as the specific material- and process-related knowledge) depends on its own costs/payments for the implementation of the innovation, the market volume, the attainable market share and price, as well as the existence of material applications which allow for large industrial scales and experience effects. It is difficult to assess as the following monetary appraisal also indicates. Overall, however, the strategic oriented non-monetary appraisal draws a relatively positive picture of powder-metallurgically produced AMCs' potential of gaining competitive advantages [5].

The *monetary appraisal* has to refer to the costs and revenues (or the corresponding payments) caused by a new technology in its entire life cycle consisting of phases, such as development, manufacturing, usage, and recycling. However, for comparing commercial potentials of powder-metallurgically produced AMCs and reference materials, it is sufficient to evaluate expected differences in costs and revenues which are relevant for decisions about the technologies. Due to a limited data base of relevant development and manufacturing costs, following considerations focus on monetary effects of improved material properties and, therefore, on their (additional) monetary benefit (for an user like an airline which might be willing to pay higher prices for aircrafts and their components). An additional monetary benefit (of usage) can be interpreted as an upper limit for higher costs of development and manufacturing of powder-metallurgically produced AMCs.

For monetary appraisal, a long-term, multiperiod perspective and the usage of dynamic methods of investment appraisal that consider time value of money have to be recommended. Here, the net present value-method is chosen [11,44] for calculating (the positive) monetary effects of weight-reduction. As outlined before, it can be assumed that an increase of 50% in yield strength enables savings in material quantities of approx. 50% and, thus, weight reductions of the same ratio [5,25]. It can be stated that the improved yield strength of the material allows for achieving other user requirements regarding quality (e.g., stability and durability) of a specific aircraft component with a reduced material thickness [25]. In the following, calculations of monetary effects of weight reduction will be demonstrated. Afterwards, some constraints will be outlined.

The two addressed (alternative) effects of weight reduction (of materials, components and, finally, aircrafts) in the usage phase are (i) revenue effects because of higher payloads, and (ii) fuel and corresponding cost saving effects when payload is the same [45]. Calculations are demonstrated by taking a Boeing 747–400 as an example [5]. This model of aircraft consists of, amongst others, 66,150 kg aluminum [46]. Assuming that 25% of this material can be substituted by the innovative AMCs with 50% less weight, a weight reduction of 8268.75 kg can be achieved. For calculating revenue effects of higher payloads, for civil aircraft industry, the monetary benefit of 1 kg weight saving is estimated as €($)100–500 for the time of material usage [35,47]. In sum, monetary benefit for the whole aircraft amounts to €826,875 (8268.75 kg·€100/kg) up to €4,134,375 (=8268.75 kg·€500/kg). These amounts can be interpreted as net present values if the monetary benefit refers to the entire life cycle of the aircraft (approx. 20–30 years) [48].

Alternatively, fuel and corresponding cost-saving effects can be calculated based on the following data:

- estimated fuel saving [49]:
  airplane (short distance): 117–134 kg kerosene p.a. per kg weight reduction
  airplane (long distance): 172–212 kg kerosene p.a. per kg weight reduction
- 1 kg kerosene = approx. 1.25 L kerosene (density: approx. 0.75–0.84 kg/L [50])
- price of kerosene: €1.50 per gallon/€0.40 per liter (1 gallon = 3.78541 L) [51]

Assuming an average fuel saving of 150 kg kerosene p.a. per kg weight reduction, fuel saving per year amounts to 1,240,312.50 kg (approx. 1,550,390.625 L) of kerosene (=150 kg/kg·8268.75 kg). It results in a cost-saving effect of €620,156.25 p.a. If the life cycle of the aircraft (and its components) spans 20 years, the monetary benefit for an airline company can be calculated as a net present value (using a discount rate of 4% based on a 'weighted average cost of capital' with typical capital structures and interest ratios) [44]:

$$620,156.25 \times \frac{1.04^{20} - 1}{1.04^{20} \times 0.04} = 8,428,125.82$$

This amount (as well as the alternatively calculated values of €826,875–€4,134,375) represents the monetary benefit of weight reduction and can be interpreted as an upper limit for costs of weight reduction (especially additional costs of development and manufacturing of powder-metallurgically produced AMCs compared to development and manufacturing costs of hitherto applied materials). If the monetary benefit exceeds the costs of weight reduction, an airline company would prefer to buy an aircraft with AMC components and would be willing to pay an additional price of max. €8,428,125.82 (average price of a Boeing 747-8 as of January 2018 was about $402.9 million [52]). This calculation implies a price of approx. €1019 (=€8,428,125.82/8268.75 kg) per kg weight reduction (which exceeds the above mentioned range of €($)100–500). It could be derived that the price of 1 kg AMC can be higher by this amount than 2 kg of a hitherto used conventional material. However, it has to be considered that this value includes not only material costs (which are much lower for commercial AMCs) but all switching costs caused by the substitution of the material, such as the costs of development or re-design of components, manufacturing processes, and additional manufacturing costs in the entire value chain (consisting of material manufacturer, aircraft component manufacturer(s), and aircraft manufacturer).

The calculations are faced by some constraints. They are based on uncertain data. Calculated values can change considerably depending on changing assumptions regarding fuel price, discount rate, aircraft life cycle, and flight distance during aircraft life [35,45]. For considering data uncertainty, especially the determination of factors with a strong influence on results, as well as critical values of influencing factors, sensitivity analyses can be conducted [44]. Furthermore, the monetary effects of weight reduction in the phases of development and manufacturing of AMC-components are neglected and only addressed here by the considerations regarding the upper limit for costs of weight reduction; they should be elaborated and analyzed in detail [42]. For example, monetary consequences of a changed geometry of components for aircraft construction could be considered as additional costs of the development and manufacturing phase. Additionally, a reduced material cross-section because of the higher yield strength potentially causes lower material quantities which are needed for producing a specific number of aircraft components [25]. Resulting cost-saving effects regarding the material costs should also be addressed by further analyses.

6. Conclusions

In this paper, the concept of an integrated technology, user, and market analysis and forecast has been presented and applied to the case of particle-reinforced aluminum matrix composites (AMCs) and their usage for manufacturing aircraft components. One the one hand, the results show the technological as well as economic potential of this innovative material. On the other hand, the principal applicability of the concept in early life cycle stages as well as its basic advantage—the "merging" of technological and economic analyses and forecasts—has been demonstrated.

However, there is considerable need for further research and development activities and results: Concerning the technological perspective, the degree of maturity of particle-reinforced aluminum matrix composites (AMCs), and the corresponding manufacturing processes has to be enhanced enabling the transfer into the industrial context. In parallel, especially the economic elements of an integrated technology, user, and market analysis and forecast have to be extended and refined in order to show a more complete picture of the economic potential of AMCs. Finally, the instrument of an integrated technology, user, and market analysis and forecast, itself, has to be further elaborated (e.g., by including more single analyzing and forecasting techniques), applied to further cases, and validated.

Acknowledgments: The authors gratefully thank the Deutsche Forschungsgemeinschaft (DFG) for funding this work within the Collaborative Research Center SFB 692.

Author Contributions: Anja Schmidt and Uwe Götze characterized the integrated methodology and the results of the user and market analysis and forecast, as well as of the technology appraisal. Steve Siebeck is responsible for the production and characterization of the aluminum matrix composites. Daisy Nestler and Guntram Wagner accompanied the material-technical part of the project and the manuscript with their professional expertise. All authors were involved in the preparation of the final version of the manuscript.

Conflicts of Interest: The authors declare no conflict of interest.

References

1. Chawla, N.; Shen, Y.L. Mechanical behavior of particle reinforced metal matrix composites. *Adv. Eng. Mater.* **2001**, *3*, 357–370. [CrossRef]

2. Beffort, O.; Long, S.; Cayron, C.; Kuebler, J.; Buffat, P.A. Alloying effects on microstructure and mechanical properties of high volume fraction SiC-particle reinforced Al-MMCs made by squeeze casting infiltration. *Compos. Sci. Technol.* **2007**, *67*, 737–745. [CrossRef]

3. Bandarian, R. Evaluation of Commercial Potential of a new Technology at the early Stage of Development with Fuzzy Logic. *J. Technol. Manag. Innov.* **2007**, *2*, 73–85.

4. Götze, U.; Schmidt, A. Innovation Control—Framework, Methods, and Applications. In *Managerial Challenges of the Contemporary Society*; Nistor, R., Zaharie, M., Gavrea, C., Eds.; "Babes-Bolyai" University of Cluj-Napoca: Cluj-Napoca, Romania, 2013; Volume 5, pp. 100–105.

5. Götze, U.; Schmidt, A.; Herold, F.; Nestler, D.; Siebeck, S. Methodik zur Analyse, Prognose und Bewertung von innovativen Werkstoffen am Beispiel von partikelverstärkten Aluminiummatrix-Verbundwerkstoffen (AMCs). In *Vorausschau und Technologieplanung zum 11. Symposium für Vorausschau und Technologieplanung, 29. und 30. Oktober 2015 in Berlin*; Gausemeier, J., Ed.; Heinz Nixdorf Institut, Universität Paderborn: Paderborn, Germany, 2015; pp. 221–241.

6. Nestler, D.; Siebeck, S.; Podlesak, H.; Wagner, S.; Hockauf, M.; Wielage, B. Powder Metallurgy of Particle-Reinforced Aluminium Matrix Composites (AMC) by Means of High-Energy Ball Milling. In *Integrated Systems, Design and Technology 2010*; Fathi, M., Holland, A., Ansari, F., Weber, C., Eds.; Springer: Berlin/Heidelberg, Germany, 2011; pp. 93–107.

7. Siebeck, S.; Nestler, D.; Wielage, B. Producing a particle-reinforced AlCuMgMn alloy by means of mechanical alloying. *Mater. Werkst.* **2012**, *43*, 567–571. [CrossRef]

8. Podlesak, H.; Siebeck, S.; Mücklich, S.; Hockauf, M.; Meyer, L.; Wielage, B.; Weber, D. Powder metallurgical fabrication of SiC and Al₂O₃ reinforced Al-Cu. *Mater. Werkst.* **2009**, *40*, 500–505. [CrossRef]

9. Siebeck, S.; Nestler, D.; Wielage, B. Hochenergie-Kugelmahlen zur Herstellung partikelverstärkter AMCs mit hochfester, maßgeschneiderter Aluminiummatrix. In *Verbundwerkstoffe, Tagungsband*; Wanner, A., Ed.; Conventus Congressmanagement & Marketing: Jena, Germany, 2013; pp. 115–121.

10. Wagner, S.; Siebeck, S.; Hockauf, M.; Nestler, D.; Podlesak, H.; Wielage, B.; Wagner, M.F.-X. Effect of SiC-Reinforcement and Equal-Channel Angular Pressing on Microstructure and Mechanical Properties of AA2017. *Adv. Eng. Mater.* **2012**, *14*, 388–393. [CrossRef]

11. Götze, U.; Hertel, A.; Schmidt, A.; Päßler, E.; Kaufmann, J. Integrated Framework for Life Cycle-Oriented Evaluation of Product and Process Technologies: Conceptual Design and Case Study. In *Technology and Manufacturing Process Selection: The Product Life Cycle Perspective*; Henriques, E., Pecas, P., Silva, A., Eds.; Springer: London, UK, 2014; pp. 193–215.

12. Drossel, G. *Umformen von Aluminium-Werkstoffen, Gießen von Aluminium-Teilen, Oberflächenbehandlung von Aluminium, Recycling und Ökologie*, 15th ed.; Aluminium-Verlag: Düsseldorf, Germany, 1999.

13. Cheng, N.P.; Zeng, S.M.; Liu, Z.Y. Preparation, microstructures and deformation behavior of SiCP/6066Al composites produced by PM route. *J. Mater. Process. Technol.* **2008**, *202*, 27–40. [CrossRef]

14. Shorowordi, K.M.; Laoui, T.; Haseeb, A.; Celis, J.P.; Froyen, L. Microstructure and interface characteristics of B_4C, SiC and Al_2O_3 reinforced Al matrix composites: A comparative study. *J. Mater. Process. Technol.* **2003**, *142*, 738–743. [CrossRef]

15. Torralba, J.M.; Da Costa, C.E.; Velasco, F. P/M aluminum matrix composites: An overview. *J. Mater. Process. Technol.* **2003**, *133*, 203–206. [CrossRef]

16. Kudashov, D. Oxiddispersionsgehärtete Kupferlegierungen mit Nanoskaligem Gefüge. Ph.D. Thesis, TU Bergakademie Freiberg, Freiberg, Germany, 2003.

17. Kaufman, J.G. *Properties of Aluminum Alloys. Tensile, Creep, and Fatigue Data at High and Low Temperatures*; ASM International; Aluminum Association: Materials Park, OH, USA; Washington, DC, USA, 1999.

18. Li, C.; Ellyin, F.; Koh, S.; Oh, S.J. Influence of porosity on fatigue resistance of cast SiC particulate-reinforced Al–Si alloy composite. *Mater. Sci. Eng. A* **2000**, *276*, 218–225. [CrossRef]

19. Wolf, M.; Wagner, G.; Eifler, D. Fatigue and Fracture Behavior of MMC in the HCF- and VHCF-Regime. *MSF* **2014**, *783–786*, 1597–1602. [CrossRef]

20. Wolf, M.; Wagner, G.; Eifler, D. Ultrasonic Fatigue of Aluminum Matrix Composites (AMC) in the VHCF-Regime. In *Supplemental Proceedings//TMS 2012 141st Annual Meeting & Exhibition*; John Wiley & Sons, Inc.: Hoboken, NJ, USA, 2012; pp. 847–853.

21. Hesse, W. *Aluminium-Werkstoff-Datenblätter/Aluminium Material Data Sheets*, 7th ed.; DIN, Deutsches Institut für Normung e.V., Ed.; Beuth: Berlin, Germany, 2016.

22. Saha, P.K. *Aerospace Manufacturing Processes*; CRC Press, Taylor & Francis Group: Boca Raton, FL, USA, 2017.

23. Dursun, T.; Soutis, C. Recent developments in advanced aircraft aluminum alloys. *Mater. Des.* **2014**, *56*, 862–871. [CrossRef]

24. Von Hippel, E. Lead Users: A Source of Novel Product Concepts. *Manag. Sci.* **1986**, *32*, 791–805. [CrossRef]

25. Herold, F.; Schmidt, A.; Frint, P.; Götze, U.; Wagner, M.F.-X. Technical-economic evaluation of severe plastic deformation processing technologies—Methodology and use case of lever-arm-shaped aircraft lightweight components. *Int. J. Adv. Manuf. Technol.* **2017**, 1–14. [CrossRef]

26. Porter, M.E. *Competitive Strategy. Techniques for Analyzing Industries and Competitors*; Free Press: New York, NY, USA, 2004.

27. Airbus. Annual Report 2016. Available online: http://company.airbus.com/investors/Annual-reports-and-registration-documents.html (accessed on 29 December 2017).

28. Boeing. Annual Report 2016. Available online: http://s2.q4cdn.com/661678649/files/doc_financials/annual/2016/2016-Annual-Report.pdf (accessed on 29 December 2017).

29. Airbus. Annual Report 2015. Available online: http://company.airbus.com/investors/Annual-reports-and-registration-documents.html (accessed on 2 February 2018).

30. Boeing. Annual Report 2015. Available online: http://s2.q4cdn.com/661678649/files/doc_financials/annual/2015/2015-Annual-Report.pdf (accessed on 2 February 2018).

31. Statista. Branchenreport Luftfahrt, 2017. Available online: https://de.statista.com/statistik/studie/id/1910/dokument/branchenreport-luftfahrt/ (accessed on 27 December 2017).

32. Kundu, A.K. *Aircraft Design*; Cambridge University Press: New York, NY, USA, 2010.

33. European Aviation Safety Agency (EASA). Regulations. Available online: https://www.easa.europa.eu/regulations (accessed on 1 February 2018).

34. MatWeb. Material Property Data. MatWeb Data Suppliers. Available online: http://www.matweb.com/reference/suppliers.aspx (accessed on 2 February 2018).

35. Ashby, M.F. *Materials Selection in Mechanical Design*, 5th ed.; Butterworth-Heinemann: Amsterdam, The Netherlands, 2017.

36. Ribeiro, I.; Kaufmann, J.; Schmidt, A.; Peças, P.; Henriques, E.; Götze, U. Fostering selection of sustainable manufacturing technologies—A case study involving product design, supply chain and life cycle performance. *J. Clean. Prod.* **2016**, *112*, 3306–3319. [CrossRef]

37. Singh, J.; Chauhan, A. Characterization of hybrid aluminum matrix composites for advanced applications—A review. *J. Mater. Res. Technol.* **2016**, *5*, 159–169. [CrossRef]

38. Mavhungu, S.T.; Akinlabi, E.T.; Onitiri, M.A.; Varachia, F.M. Aluminum Matrix Composites for Industrial Use: Advances and Trends. *Procedia Manuf.* **2017**, *7*, 178–182. [CrossRef]

39. Zönnchen, S.; Götze, U. Bewertung des kommerziellen Potenzials neuartiger Werkstoffe—Methodische Ansätze am Beispiel funktionalisierter Kohlenstofffaserwerkstoffe. In *Tagungsband zum 17. Werkstofftechnischen Kolloquium in Chemnitz, 11. und 12. September 2014*; Wielage, B., Ed.; Technische Universität Chemnitz: Chemnitz, Germany, 2014; pp. 217–232.

40. Barney, J. Firm Resources and Sustained Competitive Advantage. *J. Manag.* **1991**, *17*, 99–120. [CrossRef]

41. Barney, J.B. *Gaining and Sustaining Competitive Advantage*, 4th ed.; Pearson: Harlow, UK, 2011.

42. Schubert, A.; Götze, U.; Hackert-Oschätzchen, M.; Lehnert, N.; Herold, F.; Meichsner, G.; Schmidt, A. Evaluation of the Technical-Economic Potential of Particle-Reinforced Aluminum Matrix Composites and Electrochemical Machining. In *Tagungsband zum 18. Werkstofftechnischen Kolloquium, 10. und 11. März 2016 in Chemnitz*; Lampke, T., Wagner, G., Wagner, M.F.-X., Eds.; Technische Universität Chemnitz: Chemnitz, Germany, 2016; pp. 600–611.

43. CMT. Continuous Improvement—Research and Development. Available online: http://www.cmt-ltd.com/continuous-improvement-research-and-development (accessed on 2 February 2018).

44. Götze, U.; Northcott, D.; Schuster, P. *Investment Appraisal: Methods and Models*, 2nd ed.; Springer: Berlin, Germany, 2015.

45. Kaufmann, M.; Zenkert, D.; Mattei, C. Cost optimization of composite aircraft structures including variable laminate qualities. *Compos. Sci. Technol.* **2008**, *68*, 2748–2754. [CrossRef]

46. Hill, K. Extreme Engineering: The Boeing 747, 2011. Available online: https://sciencebasedlife.wordpress.com/2011/07/25/extreme-engineering-the-boeing-747/ (accessed on 27 December 2017).

47. Reuter, M. *Methodik der Werkstoffauswahl*, 2nd ed.; Hanser: München, Germany, 2014.

48. Franz, K.; Hörnschemeyer, R.; Ewert, A.; Fromhold-Eisebith, M.; Böckmann, M.G.; Schmitt, R.; Petzoldt, K.; Schneider, C.; Heller, J.E.; Feldhusen, J.; et al. Life Cycle Engineering in Preliminary Aircraft Design. Leveraging technology for a sustainable world. In Proceedings of the 19th CIRP Conference on Life Cycle Engineering, University of California at Berkeley, Berkeley, CA, USA, 23–25 May 2012; Dornfeld, D.A., Linke, B.S., Eds.; Springer: Heidelberg, Germany, 2012; pp. 473–478.

49. Helms, H.; Lambrecht, U.; Hanusch, J. Energieeffizienz im Verkehr. In *Energieeffizienz. Ein Lehr- und Handbuch*; Pehnt, M., Ed.; Springer: Berlin, Germany, 2010; pp. 309–329.

50. Chemeurope.com. Jet Fuel. Available online: http://www.chemeurope.com/en/encyclopedia/Jet_fuel.html (accessed on 1 February 2018).

51. IndexMundi. Jet Fuel Monthly Price. Available online: https://www.indexmundi.com/commodities/?commodity=jet-fuel&currency=eur (accessed on 1 February 2018).

52. Statista. Average Prices for Boeing Aircraft as of January 2018, by Type (in Million U.S. Dollars). Available online: https://www.statista.com/statistics/273941/prices-of-boeing-aircraft-by-type/ (accessed on 19 February 2018).

MDPI

St. Alban-Anlage 66

4052 Basel

Switzerland

Tel. +41 61 683 77 34

Fax +41 61 302 89 18

www.mdpi.com

Metals Editorial Office

E-mail: metals@mdpi.com

www.mdpi.com/journal/metals

www.ingramcontent.com/pod-product-compliance
Lightning Source LLC
Chambersburg PA
CBHW051853210326
41597CB00033B/5880

* 9 7 8 3 0 3 8 9 7 1 9 6 2 *